Therapeutic Perspectives of Tea Compounds

Phytochemicals have a wide variety of biological activities, including broad antimicrobial activities. Medicinal plant-derived compounds tend to be lowly toxic. More than 50% of medical drugs used in western countries are derived from plant-based compounds. The application of naturally derived, especially plant-based compounds for supportive and auxiliary treatments is attracting increasing attention. Tea (*Camelia sinensis*) could be a promising candidate for the prevention and treatment of several diseases. The use of tea constituents over synthetic drugs is an exciting treatment option since they are safer and a higher dose is feasible.

Although effective vaccines have found clinical applications, they are not available in many countries yet. Thus, simple and cheap measures which are easily available for everyone to help mitigate COVID-19 are needed. Dietary bioactive compounds could be alternative nutritional approaches in combating COVID-19 infection. The consumption of these nutrients in daily diets can support existing therapies, upcoming vaccines, and drugs by enhancing their efficacy. Tea is the most widely consumed beverage in the world, and developing antiviral molecules from the same is an exciting idea that has advantages over synthetic toxic antiviral drugs concerning higher dosages. Antiviral properties of the tea compounds against COVID-19 together with other beneficial health effects of these compounds could be advancement in treatment to this latest pandemic.

This book presents the chemical composition of tea and the beneficial health effects of the tea compounds focusing on the various antiviral properties of these compounds and their effectiveness in suppressing the propagation of SARS-CoV-2. The book brings evidence for the effectiveness of these compounds, and compares them with the most potent available anti-COVID-19 drugs, supporting the development of promising anti-COVID-19 drugs from tea.

Medicinal Plants and Natural Products for Human Health

Series Editor:
Christophe Wiart

We are at a point in history where the global population is increasingly interested in medicinal plants, natural products and herbalism. This interest has accelerated with COVID-19 and long COVID symptoms. Rediscovery of the connection between plants and health is responsible for a new generation of botanical therapeutics that include plant-derived pharmaceuticals, multicomponent botanical drugs, dietary supplements, functional foods and plant-produced recombinant proteins. Many of these products can complement conventional pharmaceuticals in the treatment, prevention and diagnosis of diseases. This series is a critical reference for anyone involved in the discovery of biopharmaceuticals for the improvement of human health.

Alternative Medicines for Diabetes Management: Advances in Pharmacognosy and Medicinal Chemistry
Edited by: Varma H. Rambaran, Nalini K. Singh

Therapeutic Perspectives of Tea Compounds: Potential Applications against COVID-19
Authored by: Farnoosh Dairpoosh and Kianoosh Dairpoosh

Therapeutic Perspectives of Tea Compounds

Potential Applications Against COVID-19

Farnoosh Dairpoosh
and
Kianoosh Dairpoosh

CRC Press
Taylor & Francis Group
Boca Raton London New York

CRC Press is an imprint of the
Taylor & Francis Group, an **informa** business

First edition published in 2023
by CRC Press
6000 Broken Sound Parkway NW, Suite 300, Boca Raton, FL 33487-2742

and by CRC Press
4 Park Square, Milton Park, Abingdon, Oxon, OX14 4RN

CRC Press is an imprint of Taylor & Francis Group, LLC

ISBN: 9781032463124 (hbk)
ISBN: 9781032466446 (pbk)
ISBN: 9781003382652 (ebk)

DOI: 10.1201/9781003382652

Typeset in Garamond
by codeMantra

Contents

Author

Dr. Farnoosh Dairpoosh earned a BSc in Cell and Molecular Biology-Microbiology and an MSc in Biology in Tehran, Iran. She has great interest in analyzing metabolites and bioactive compounds. She was awarded a full scholarship from Jacobs University (the former International University Bremen), and she moved to Germany and earned a PhD in Analytical Chemistry-Natural Products from that university in 2011. During her PhD studies, she worked on the identification and the profiling of polyphenols in the European Diet. She has identified and reported the presence of several polyphenolic compounds not previously reported in fruits and vegetables, in the literature. She completed her post-doctoral fellowship in Analytical Biochemistry at the Technical Biochemistry Institute at Saarland University in Saarbrücken, Germany. She has worked as a senior scientist both in academia and in the pharma-biotech industry. Her research interests revolve around bioactive natural compounds, functional foods, biopharmaceuticals, and metabolic disorders.

Kianoosh Dairpoosh earned a BSc in Chemistry from Kharazmi University, Tehran, Iran. She earned an MSc in Nanomolecular Science from Jacobs University, School of Engineering and Science, Bremen, Germany, in 2010. She has worked extensively on the investigation of polyphenols in green and black teas. Since then, she has worked as a research scientist in the biotech industry. Her research interests revolve around nutrition, bioactive natural compounds, and functional foods.

Chapter 1

Introduction

Tea is an infusion of *Camellia sinensis* leaves (Figure 1), which is the most consumed beverage in the world, except for water. Tea is generally divided into three types:

- Green tea
- Black tea
- Oolong tea

All of these types of tea are produced from tea (*Camellia sinensis*) leaves. Green tea is produced by completely inactivating polyphenol oxidases (PPOs) in tea leaves by exposure to steam vapour for 30 seconds, after which they are rolled (crushed) by hand or machine. Green tea is thus known as unfermented tea since this inactivation of the enzyme ensures that fermentation does not take place. Black tea is produced by treating fresh tea leaves at a temperature of 28°C for 18 hours, thus reducing the water content by half (Kuroda and Hara, 2004).

The tea leaves are then rolled by machine and fermented at 25°C to 30°C for 2–3 hours. Black tea thus produced according to these processes is called fermented tea. However, this fermentation process is not due to microorganisms but

DOI: 10.1201/9781003382652-1

Figure 1   The tea plant is classified as *Camellia sinensis*.

is brought about by the oxidation of catechins by oxidative enzymes mainly PPOs that are present in tea leaves. During the process of fermentation, the colour of the tea changes to a reddish tone. Oolong tea is produced by a process in which fresh tea leaves are half fermented by enzymes present in tea leaves. Therefore, oolong tea is called half fermented (Kuroda and Hara, 2004).

1.1 Manufacturing of Black Tea

The manufacture of tea includes the collection and drying of tea leaves which plays an important role in the quality and flavour of the product. Tea manufacturing practices are very important because of their effects on the quality, taste, and price of the product. Manufacturing of tea includes the following process:

1. Partial removal of moisture (withering)
2. Leaf disruption into small pieces (maceration)
3. Quality development by exposure to air (fermentation)
4. Completion of moisture removal (drying)
5. Sieving into size fractions with fibre removal (sorting)

Withering is the first step in which chemical changes take place and moisture levels are decreased from 75%–80% to 55%–70%. Usually, this process takes about 12–16 hours; meanwhile, the air is blowing through the leaves to remove moisture. Having a constant temperature is important because overheating leads to an unfavourable chemical reaction which affects the quality of tea.

Maceration is the stage in which reducing the size of leaf particles and cell disruption happens. Cell disruption results in bringing polyphenolic compounds (flavan-3-ol) into the vacuole and in contact with PPOs in the cytoplasm and activating many other enzymes simultaneously (Crozier *et al.*, 2006). PPOs are copper-containing enzymes that are nearly ubiquitous among plants (Mayer, 2006). In a particular plant, PPO activity varies from one organ to another and varies inside an organ, depending on the tissue considered. PPOs have been found in different cell fractions, in organelles (chloroplasts and, more precisely, in thylakoids, mitochondria, and peroxisomes) where the enzymes are tightly bound to membranes and in the soluble fraction of the cell (Nicolas *et al.*, 2003). PPOs have broad substrate specificity and are associated with the browning of plant materials. Because PPOs are often induced by wounding or pathogen attacks, they are most generally believed to play important roles in plant defence responses.

During the fermentation stage, polyphenols are oxidized to form the characteristic compounds of black tea. The biochemical and chemical reactions taking place during this stage are exothermic. The temperature must be controlled to prevent unfavourable secondary reactions. Temperature control is provided by passing air through the macerated leaves. Also, it results in the provision of oxygen needed for the oxidation reactions. By controlling the temperature, leaves are treated with air quantities in vast excess of that needed for the fermentation process only. The duration of the fermentation stage is around 60 minutes but can be much shorter or longer depending on the factory machinery used to disrupt the leaf and the ambient temperature (Hampton, 1992).

Drying is the next step after the fermentation process. In this stage, warm air flows are used to eliminate moisture. Skilful drying in which moisture is decreased to 2.5; 2.5%–3.5% is important for the quality of the product. Also, the temperature needs to be controlled because too high temperature denatures and inactivates the enzymes in black tea (Hampton, 1992).

Sorting is used to separate the different-sized tea particles into even-sized groups and to remove the fibres and stalk in the tea. The stalk and fibre are removed by electrostatically charged rollers, which attract the fibre and stalk preferentially. After sorting, the tea is packed and transported to the market (Hampton, 1992).

1.2 Manufacturing of Green Tea

There are basically two types of green tea (Takeo, 1992). The Japanese type of green tea consists of shade-grown hybrid leaves with low flavan-3-ol levels and high amino acid content including theanine. After harvesting, the leaves are steamed rapidly to inhibit polyphenoloxidase and other enzymes.

The Chinese type of green tea leaves are exposed to dry heat (firing) rather than steaming, therefore less inhibition of the PPO activity and some transformations of the flavan-3-ols occur (Fan *et al.*, 1999). In green tea manufacturing, fresh green tea leaves are wilted quickly at high heat; thus, cell enzymes are immediately denatured and no fermentation is taking place (Huang, 2001).

1.3 Manufacturing of Oolong Tea

Oolong tea leaves are wilted but not rolled and are fermented partially. By adjusting the processing conditions, the fermentation of oolong tea varies from 15% to 60% (Huang, 2001). Approximately 30% of catechins and 20% of proanthocyanidins are oxidized during the manufacture of oolong tea from fresh shoots, and 20% of all the flavonoids are decomposed in the drying process (Dou *et al.*, 2007).

1.4 Quality Assessment of Green Tea and Black Tea

For quality assessment, quantitative biochemical tests were developed to monitor biochemical reactions taking place during fermentation and black tea manufacturing. The most reliable biochemical methods are based on assays of PPO activity and the catechin content of tea plants. These factors are best manifested in the theaflavin (TF) and thearubigin (TR) content in the black tea beverage. Another biochemical indicator of the quality of brewed black tea is the "volatile flavour index", based on the volatile compounds in black tea (Owuor and Langat, 1988).

   Green tea quality is primarily based on the catechin content of the leaves. Flavan-3-ols synthesized in tea leaves are the most important non-volatile constituent substrates of black tea. The flavan-3-ols provide the characteristic taste. Six major flavan-3-ols occur in green tea leaves: (+)-catechin (C), (+)-gallocatechin (GC), (−)-epicatechin (EC), (−)-epigallocatechin (EGC), (−)-epicatechin-3-gallate (ECG) and (−)-epigallocatechin-3-gallate (EGCG). The more the catechin content of the tea leaves are, the more antioxidant activity and biological properties with potential health benefits are gained. The concentration of toxic metal elements could also significantly influence the biological properties of the tea leaves and thus quality and safety (Koch *et al.*, 2018).

1.5 Factors Influencing the Quality of Tea

Together with the manufacturing practices followed, the quality of tea is also influenced by a variety of environmental and horticultural factors. Environmental factors include variations in temperature, rainfall, and the amount of sunlight. Horticultural factors include fertilization, plucking standards, and the frequency of plucking (Ellis and Nyirenda, 1995).

Chapter 2

Chemical Components of Tea

The largest component of green tea leaves is carbohydrates including cellulosic fibre. The next biggest components are insoluble proteins (Yamamoto *et al.*, 1997).

2.1 Caffeine

Caffeine is a trimethyl derivative of purine 2,6-diol and is synthesized mainly in the leaves of the tea plant (Nakabayashi *et al.*, 1991).

2.2 Amino Acids and Other Nitrogenous Compounds

About one-fifth of the total nitrogen found in a green tea infusion originates from caffeine and related compounds. Other nitrogenous compounds in the tea infusion are amino acids, amides, certain proteins, and nucleic acids. Green tea contains about 4.5%– 6.0% nitrogen and about half of it is free amino acids (Yamamoto *et al.*, 1997).

DOI: 10.1201/9781003382652-2

2.3 Vitamins

Commercial green tea leaves contain vitamin C (ascorbic acid) of about 280 mg per 100-g dried leaves. But the vitamin C content of oolong tea or black tea is less than that of green tea, since vitamin C is decomposed during the fermentation process (Yamamoto *et al.*, 1997).

2.4 Inorganic Elements

Some specific inorganic compounds in the tea plant are aluminium (Al), fluorine (F) and manganese (Mn). The levels of AI and F in tea leaves are relatively higher than in other plants. Nagata, applying 27Al NMR analysis, observed that in the tea leaf aluminium exists mainly in a chelate form (Nagata, 1990).

2.5 Carbohydrates

The total carbohydrate contained in green tea leaves is about 40%, and one-third of it is cellulosic fibre. Starch is also contained and affects the quality of green tea (Yamamoto *et al.*, 1997).

2.6 Lipids

4% of the weight of tea leaves is lipids (Chang and Chen, 1994).

2.7 Alkaloids

Tea leaves contain caffeine (1, 3, 7-trimethylxanthine) and methylxanthines, theobromine (3, 7-dimethylxanthine) and theophylline (1, 3-dimethylxanthine) (Johnson and Williamson, 2003).

2.8 Volatile Compounds

About 630 volatile compounds were reported in tea leaves. The aromatic compounds among these occur in only low levels (Chang and Chen, 1994).

2.9 Phenolic Compounds

Highly interesting components of green tea leaves are polyphenols which are usually extracted with hot water or with ethyl acetate from the aqueous phase of a tea infusion (Wright, 2005). Polyphenols can be classified as polycyclic types such as flavonoids, anthraquinones, anthocyanidins and others. The chemical structures of some of these classes are described:

2.10 Classification of Phenolic Compounds

Phenolic compounds are classified into two groups:

■ Flavonoids
■ Non-flavonoids

2.10.1 Flavonoids

The term 'flavonoid' is generally used to describe a broad collection of structurally similar natural products that include a C_6-C_3-C_6 carbon framework. Depending on the position of the linkage of the aromatic ring to the benzopyran (Figure 2) moiety, this group of natural products may be divided into three classes (Figure 3):

1. Flavonoids
2. Isoflavonoids
3. Neoflavonoids

Flavonoid compounds fall into six subclasses:

1. Flavones
2. Flavanones
3. Isoflavones
4. Flavan-3-ols
5. Flavonols
6. Anthocyanidins

Numerous structures of flavonoids are theoretically possible, based on the assumption that ten carbons of the flavonoid skeleton can be substituted by a range of different substituents, among which are hydroxyl, methoxyl, methyl, isoprenyl and

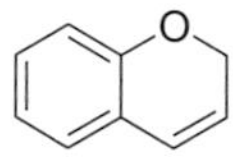

Figure 2   Benzopyran.

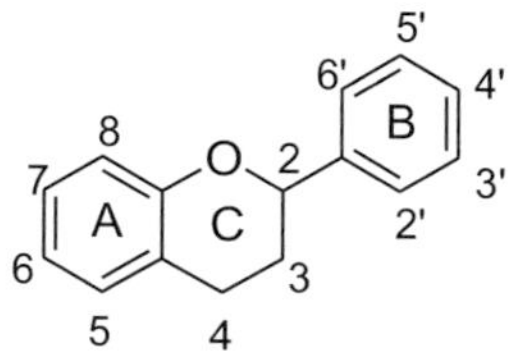

1

flavonoides

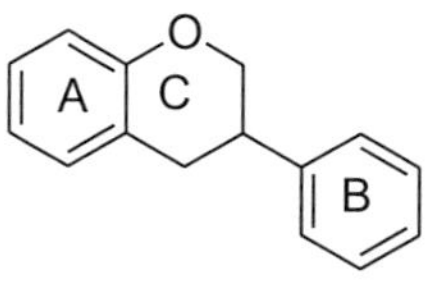

2

isoflavonoides

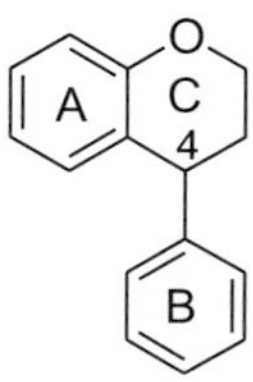

3

neoflavonoides

Figure 3   The structure of flavonoids, isoflavonoids and neoflavonoids (Grotewold, 2006).

benzyl substituents. Furthermore, each hydroxyl group and some carbons can be substituted by one or more of a range of different carbohydrates and in turn, each sugar can be acylated with a variety of different phenolic or aliphatic acids (Figure 2) (Grotewold, 2006).

Based on the degree of oxidation and saturation present in the C-ring, the flavonoids may be divided into flowing groups (Figure 4).

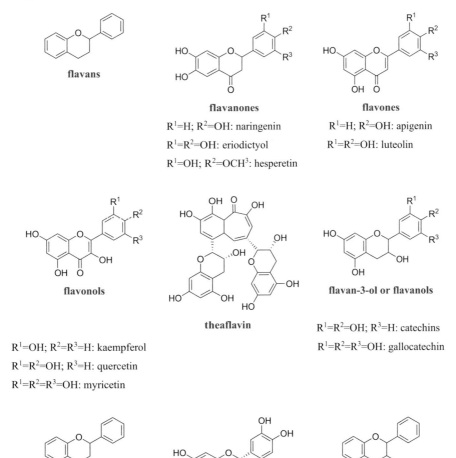

flavans

flavanones

R^1=H; R^2=OH: naringenin
R^1=R^2=OH: eriodictyol
R^1=OH; R^2=OCH^3: hesperetin

flavones

R^1=H; R^2=OH: apigenin
R^1=R^2=OH: luteolin

flavonols

R^1=OH; R^2=R^3=H: kaempferol
R^1=R^2=OH; R^3=H: quercetin
R^1=R^2=R^3=OH: myricetin

theaflavin

flavan-3-ol or flavanols

R^1=R^2=OH; R^3=H: catechins
R^1=R^2=R^3=OH: gallocatechin

flavan-4-ol

(-)-epicatechin

flavan-3,4-diol

Figure 4   Different groups and examples of flavonoids (Grotewold, 2006).

Within the classes of flavonoids, various differentiations occur based on the number and the nature of substituents attached to the rings (Cavaliere *et al.*, 2007). Flavonoids are responsible for the colouration of flowers, fruits and sometimes leaves. More than 4,000 flavonoids have been reported in nature and new flavonoids are constantly added to the list (Fabre *et al.*, 2001). Flavonoids, along with other secondary metabolites are believed to be biosynthesized for protection against UV light, pathogens and herbivores. In plants, the dominant form of flavonoids is a flavonoid glycoside. In all flavonoid glycosides, a glycosidic moiety is attached via either an O (-O-) atom or a skeletal C atom (-C-) to the skeletal C_{15} (March *et al.*, 2006). Flavonoids exist everywhere in nature; they are found in petals, and leaves and are widely distributed in the edible parts of plants (March and Miao, 2004). The flavonoid subclasses and patterns of glycosylation are strongly correlated with plant taxonomy and give rise to a wide range of chemical properties (Lin and Harnly, 2007).

2.10.1.1 Flavonols

Flavonols such as quercetin, kaempferol, myricetin and isorhamnetin are the most ubiquitous flavonoids in food which are found in vegetables such as Capers, Chives, Onions (the richest source; up to 1.2 g/kg fresh weight) and the leaves such as Lettuce, Celery and Broccoli. The glycoside concentration in the green outer leaves is about ten times higher than in the inner light-coloured leaves. In cereals such as buckwheat, beans and in fruits such as apples, apricots, grapes, plums, bilberries, blackberries, blueberries, cranberries, olive, elderberries, currants and cherries contain between five and ten different flavonol glycosides. Flavonols are also found in spices and herbs such as Dill Weed. Other dietary sources for flavonols are red wine, tea (green, black), cocoa powder, turnip (green), endive and leek (Manach *et al.*, 2004; Han *et al.*, 2007).

2.10.1.2 Flavones

Flavones are not distributed widely. The most important examples are glycosides of apigenin and luteolin (Figure 4). Flavones are mostly found in celery, parsley and some herbs (Manach *et al.*, 2004; Han *et al.*, 2007).

2.10.1.3 Flavanones

Flavones are found mostly in tomatoes and certain aromatic plants such as mint, but in high concentrations only in citrus fruits. The main aglycones are naringenin in grapefruit, hesperetin in oranges (200–600 mg/L) and eriodictyol in lemons. The solid parts of citrus fruits (the white spongy portion) and the membranous portions are high in flavanones compared to the pulp, therefore the whole fruit may contain up to five times as much as a glass of orange juice. Apigenin and luteolin are mostly found in fruits such as olives, in vegetables such as hot peppers, fresh parsley, spices and herbs such as oregano, rosemary, dry parsley and thyme (Manach *et al.*, 2004; Han *et al.*, 2007).

2.10.1.4 Isoflavones (Flavans)

Isoflavones including genistein and daidzein (Figure 5) are found in fruits such as Grapes, the seed and skin and in leguminous plants. Soya is the main source of isoflavones in the human diet. They contain three main molecules: genistein, daidzein and glycitein. The isoflavones are found as aglycones and glycosylated derivatives of the aglycones (Manach *et al.*, 2004; Han *et al.*, 2007).

2.10.1.5 Flavanols (Flavan-3-ols)

Flavanols exist in monomeric form (catechins) and the oligomeric form (proanthocyanidins (PAs)) and are found in many fruits such as apples, apricots, grapes, peaches, nectarines, pears, raisins, raspberries, cherries, blackberries, blueberries and cranberries. Other sources of flavanols are red wine, tea, chocolate, wine and cocoa. An infusion of green tea contains up to 200 mg of catechins. Black tea contains fewer monomer flavanols, which are oxidized during the fermentation of tea leaves to more complex condensed polyphenols known as theaflavins and thearubigins (Manach *et al.*, 2004; Han *et al.*, 2007).

2.10.2 Isoflavonoids

Isoflavonoids are a distinctive subclass of flavonoids. Isoflavonoids have limited distribution in the plant kingdom but their structural variations are considerably diverse. Isoflavonoids are subdivided into the following groups (Grotewold, 2006) (Figure 5).

2.10.3 Neoflavonoids

The neoflavonoids are structurally related to the flavonoids and the isoflavonoids (Figure 6).

2.10.4 Minor Flavonoids

Natural products such as chalcones and aurones also contain a C6-C3-C6 backbone and are considered to be minor flavonoids (Grotewold, 2006) (Figure 7).

2.10.5 Anthocyanidins and Anthocyanins

Anthocyanidins

They are the largest group of water-soluble pigments in the plant kingdom. They are dissolved in the vacuoles of epidermal tissues of plants which exhibit different colours. They are distinguished from other flavonoids as a separate class by their ability to exist in the flavylium ion form. Highly red, blue and purple-coloured vegetables, flowers and fruits such as

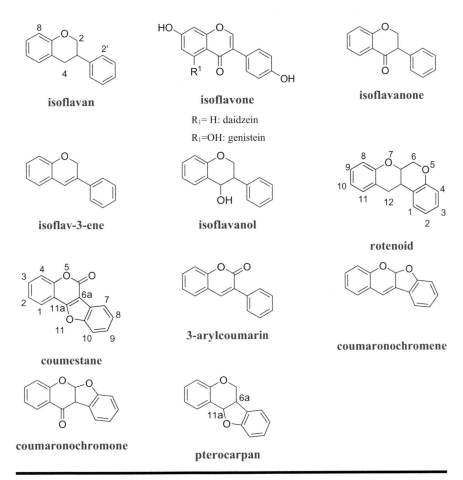

Figure 5   Different structures of isoflavonoids (Grotewold, 2006).

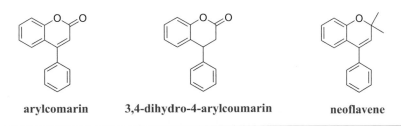

Figure 6   Different structures of neoflavonoids (Grotewold, 2006).

blueberries, cherries, raspberries, strawberries, black currants and purple grapes tend to be the major source of anthocyanidins (Mazza, 2007).

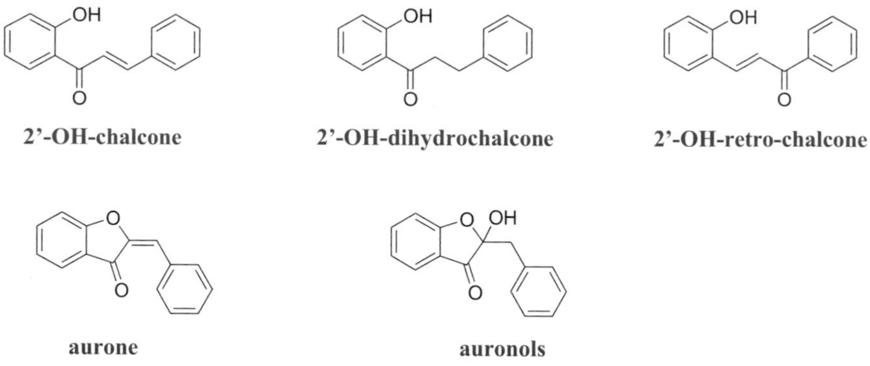

2'-OH-chalcone   2'-OH-dihydrochalcone   2'-OH-retro-chalcone

aurone   auronols

Figure 7   Different structures of minor flavonoids (Grotewold, 2006).

Colour differentiations among anthocyanidins are caused by the different substitution patterns of the B-ring of the gly-cone, the pattern of glycosylation, the degree and nature of esterification of the sugars with aliphatic or aromatic acids and by the pH, temperature, type of solvent and the presence of co-pigments (Mazza, 2007). The six anthocyanidins commonly found in plants are pelargonidin, cyanidin (the most commonly occurring anthocyanidin in nature), delphinidin, peonidin, petunidin and malvidin (Figure 8).

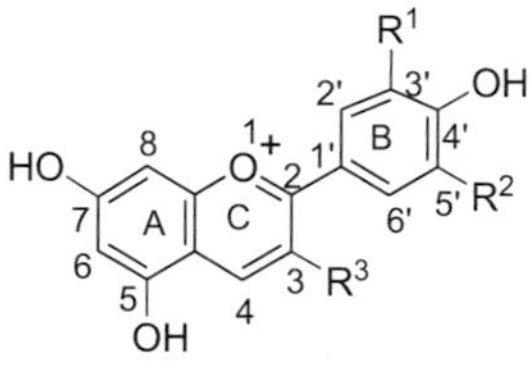

	R^1	R^2	R^3
delphinidin	OH	OH	OH
cyanidin	OH	H	OH
petunidin	OCH$_3$	OH	OH
peonidin	OCH$_3$	H	OH
malvidin	OCH$_3$	OCH$_3$	OH
pelargonidin	H	H	OH

Figure 8   Structural classification of common anthocyanidin species (Mazza, 2007).

Anthocyanins
They are the glycosylated derivative of anthocyanidins that
are responsible for the pink-red colours of most flower petals,
most red fruits (like apples) and almost all red leaves during
the autumn (Han *et al.*, 2007).

The normal tea leaves in green colour contain a low level of
anthocyanins (Hu *et al.*, 2020). Tea cultivars with purple leaves,
owing to their high content of anthocyanins, are increasingly
attracting the attention of tea growers and tea consumers.
However, the available cultivars of purple-leaf tea are limited
(Hu *et al.*, 2020).

Zijuan' (ZJ) is a tea cultivar with a purple leaf from Yunnan
province in China where the weather is hot and humid (Hu *et al.*,
2020). This tea is a rich source of anthocyanins (Hu *et al.*, 2020).

Another example is a modified form of tea named purple tea
(PT) which is derived from *Camellia sinensis* developed by the
Tea Research Foundation of Kenya (TRFK). PT is unique among
other tea in composition and contains high levels of anthocyanins
and anthocyanidins compared to normal tea (Khan *et al.*, 2018).

Black tea processing results in a significant loss of foliar
anthocyanins and green tea processing has little effect on foliar
anthocyanins. To maintain a high level of foliar anthocyanins,
fresh tea leaves with abundant anthocyanins from purple leaf
cultivars should be processed into green tea instead of black
tea (Hu *et al.*, 2020).

2.10.6 *Phenolic Acids*

The main non-flavonoid dietary phenolic compounds are the
C6–C1, C6-C3 and C6-C5 phenolic acids and their derivatives.
Two classes of phenolic acids can be recognized: (1) derivatives of benzoic acid and (2) derivatives of cinnamic acid. The
common hydroxycinnamates are p-coumaric acid, caffeic acid,
ferulic acid and sinapic acid conjugates.

Gallic acid and chlorogenic acid are examples of phenolic
acids in tea. It is contained in black tea to a larger extent than
in green tea because it is released from gallated catechins

during the fermentation procedure (Johnson and Williamson, 2003). Tea leaves may contain up to 4.5 g/kg of fresh weight (Manach *et al.*, 2004).

Figure 9 shows the structures of some phenolic acids in the plants (Johnson and Williamson, 2003; Crozier *et al.*, 2009).

2.10.7 Anthraquinones

Anthraquinones are mostly present in plants as glycosides. Anthraquinones were not detected in any vegetables or fruits, but they are known to be contained in some medicinal plants such as Aloe, Cascara Sagrada and Senna. Usually, anthraquinones are found in the form of aglycone with one or more substituent molecules. Anthraquinone (Figure 10), is

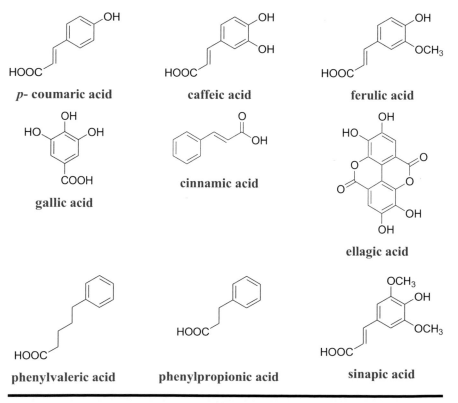

Figure 9   Chemical structures of some phenolic acids (Crozier *et al.*, 2009).

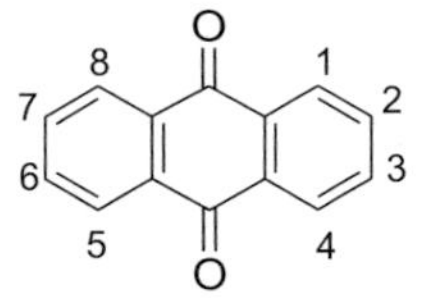

Figure 10   Anthraquinone (Urquiaga and Leighton, 2000).

a polycyclic aromatic hydrocarbon containing two opposite carbonyl groups (C=O) at the 9 and 10 positions (Urquiaga and Leighton, 2000).

2.10.8 Stilbenes

Stilbenes have a C6–C2–C6 structure, and they are produced by plants in response to disease, injury and stress. These compounds are found only in low quantities in dietary plants. An example of the stilbenes is resveratrol (a major compound in red wine) which has been well-studied in medicinal plants (Figure 11) (Manach *et al.*, 2004; Crozier *et al.*, 2009).

2.10.9 Tannins

Tannins are a group of water-soluble polyphenols having molecular weights from 500 to 3,000 which are subdivided into condensed and hydrolyzable tannins and commonly found in complexes with alkaloids, polysaccharides and proteins, particularly the latter (Han *et al.*, 2007). Tannins are responsible for the astringent taste of fruits. The level of

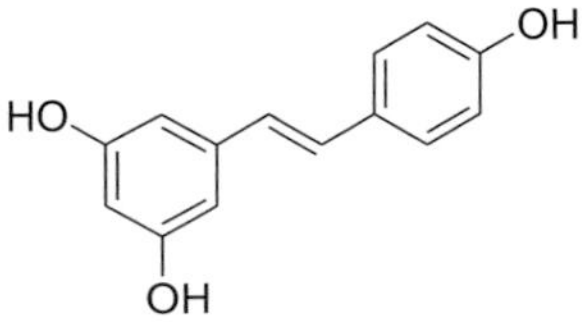

Figure 11   Resveratrol (Manach *et al.*, 2004).

astringency changes over maturation and often disappears when the fruit reaches ripeness. Polymerization of the tannins with acetaldehyde is an explanation for the reduction in the amount of tannins observed during the ripening of many types of fruits. Pas, also known as condensed tannins, have a wide range of structures and molecular weights which make it difficult to estimate their amounts in plants (Manach *et al.*, 2004; Han *et al.*, 2007). Tannins are found in fruits and vegetables such as grapes, strawberries, raspberries, pomegranates, walnut, peach, blackberry, olive, plums, chickpeas and lentils. Other sources of tannins are Haricot Bean, red wine, cocoa, chocolate, tea, cider, coffee and immature fruits. The structures of some tannins are illustrated in Figure 12 (Falcão and Araújo, 2011, 2018).

2.10.10 Lignans

The highest amount of lignans (Figure 13), which are considered dimers of coniferyl alcohol, can be found in linseed (up to 3.7 g/kg dry wt). Other cereals, grains, fruits and certain vegetables also contain traces of lignans, but concentrations in linseed are about 1,000 times as high as concentrations in other food sources (Manach *et al.*, 2004).

2.10.11 Green Tea Flavonoids

Tea leaves are rich in flavanols (flavan-3-ols) which are the largest polyphenolic component in green tea leaves (Banerjee and Chaudhuri, 2005; Sakakibara *et al.*, 2003).

The flavanols found in green tea include:

- (+)-Catechin
- (−)-Gallocatechin
- (−)-Epicatechin
- (−)-Epigallocatechin
- (−)-Epicatechin-3-gallate

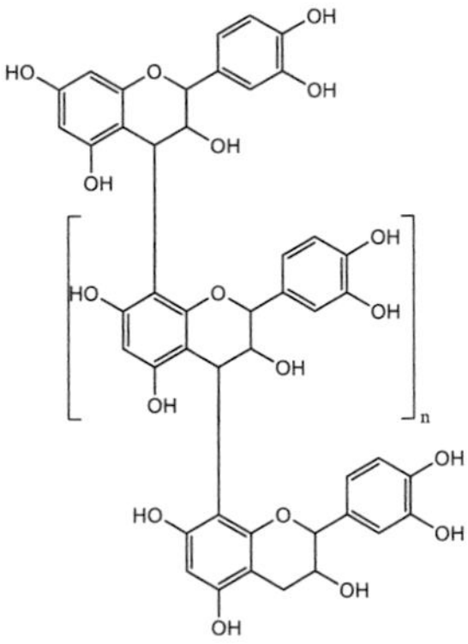

Condensed tannins

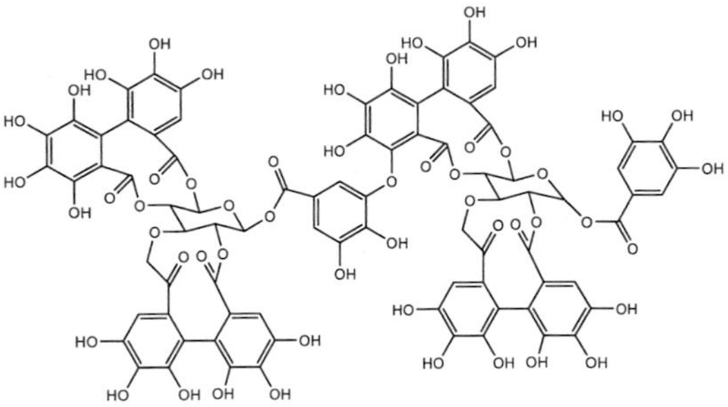

ellagitannins

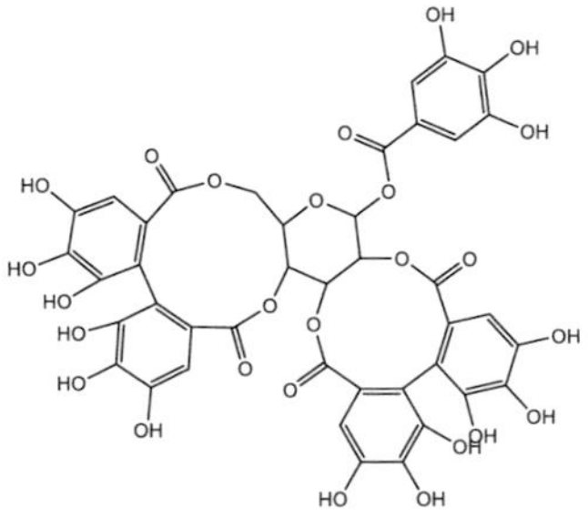

casuarictin

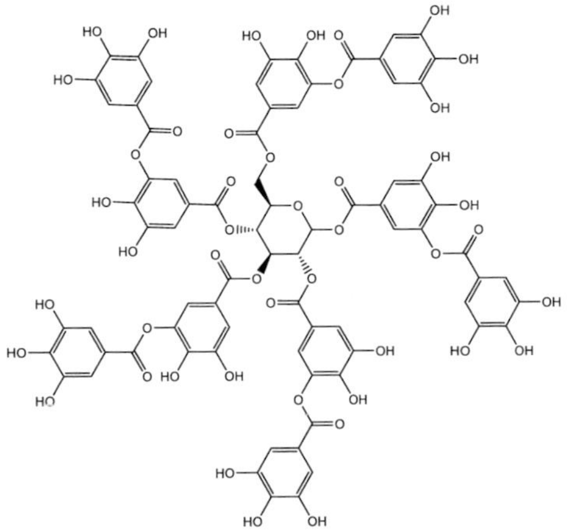

tannic acid

Figure 12   Structure of some tannins (Falcão and Araújo, 2011; Falcão and Araújo, 2018).

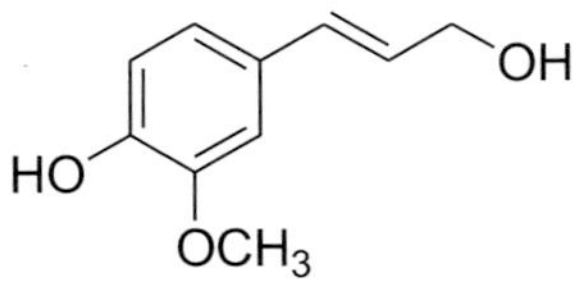

coniferyl alcohol

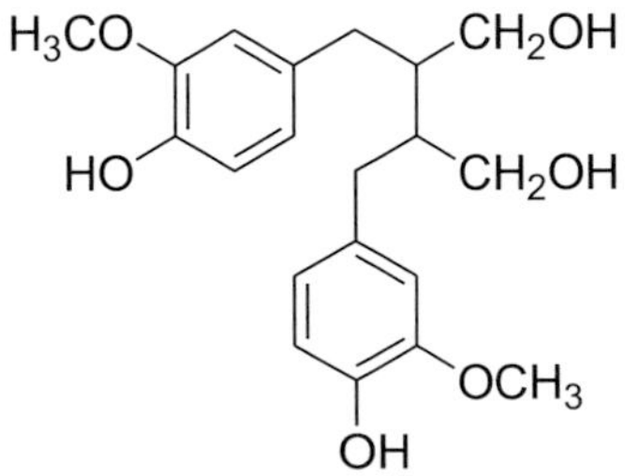

lignan

Figure 13   Chemical structure of lignan and coniferyl alcohol (PubChem).

- ■ (−)-Epigallocatechin-3-gallate
- ■ (−)-Epicatechin-3,5-digallate
- ■ 3-methyl digallate of (−) epicatechin
- ■ 3-methyl digallate of (−)-epigallocatechin

Green tea catechins and their derivatives differ depending on the species of the tea plant and the season of harvesting. Among the green tea catechins, epigallocatechin-3-gallate (also known as epigallocatechin gallate) is the largest in quantity. Figure 14 shows examples of green tea catechins and their derivates.

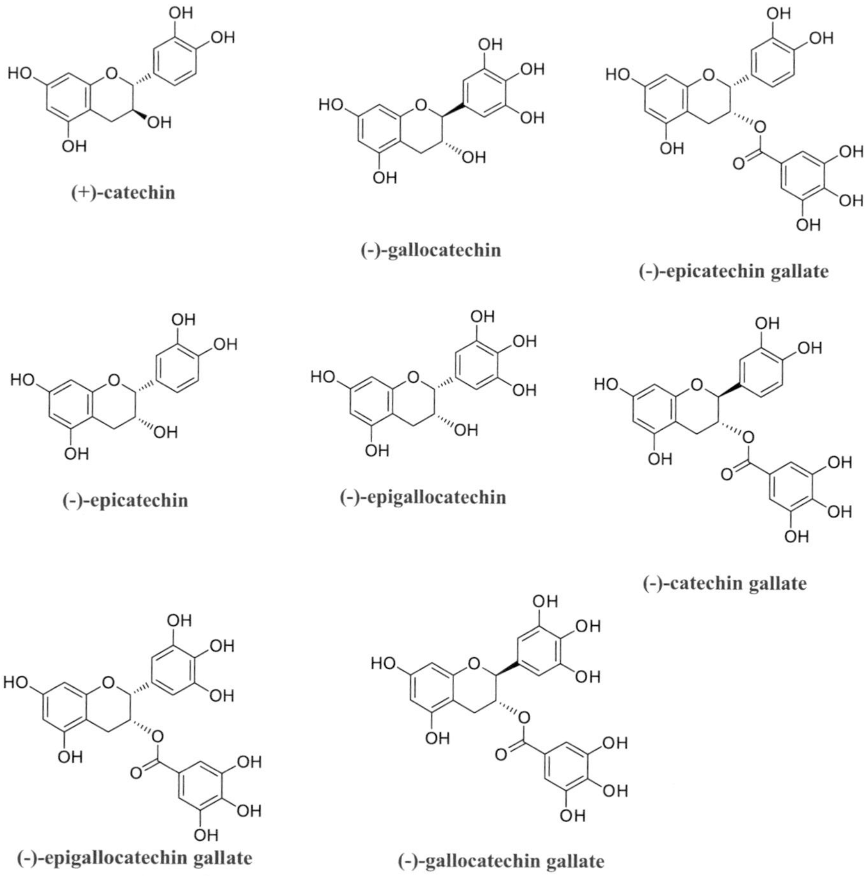

Figure 14   **Structures and examples for green tea catechin and derivatives (Sakakibara et al., 2003).**

Proanthocyanidins

Proanthocyanidins (PAs) are a group of flavonoids in fresh tea leaves. PAs are oligomeric and polymeric flavan-3-ols. Procyanidin B2 and prodelphinidin B4 are two examples of PAs in tea (Hashimoto and Ono, 2003) (Figure 15).

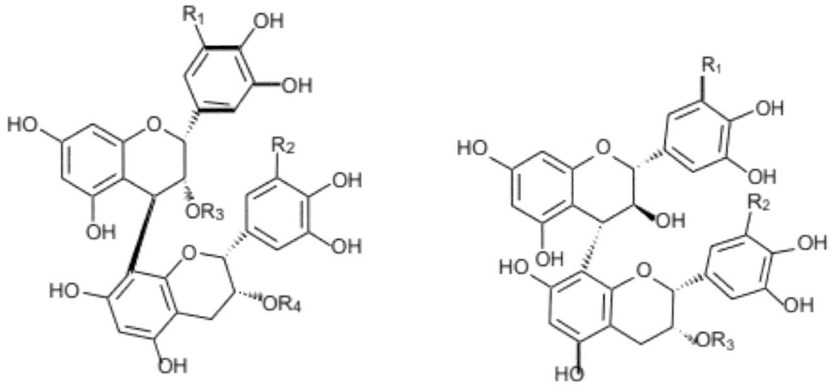

Procyanidin B-2   Procyanidin B-4

Figure 15   Structures of procyanidin B-2 and procyanidin B-4 (PubChem).

2.10.12 *Black Tea Polyphenols*

Black tea contains caffeine and flavonoids such as quercetin, rutin kaemferol and myricetin (Chang and Chen, 1994).

During the fermentation process of black tea, green tea leaves catechins by enzymatic oxidation are converted to theaflavins and thearubigins and theasinensins. They are poorly characterized chemically (Haslam, 2003; Mathew and Parpia, 1971) (Figure 16).

As is suggested in Figure 16, catechins (EGC) and (EGCG) are oxidized to ortho-quinones by polyphenol oxidase (Roberts, 1957) and the major products that are obtained after secondary oxidation are three bis-flavonols (theasinensins),

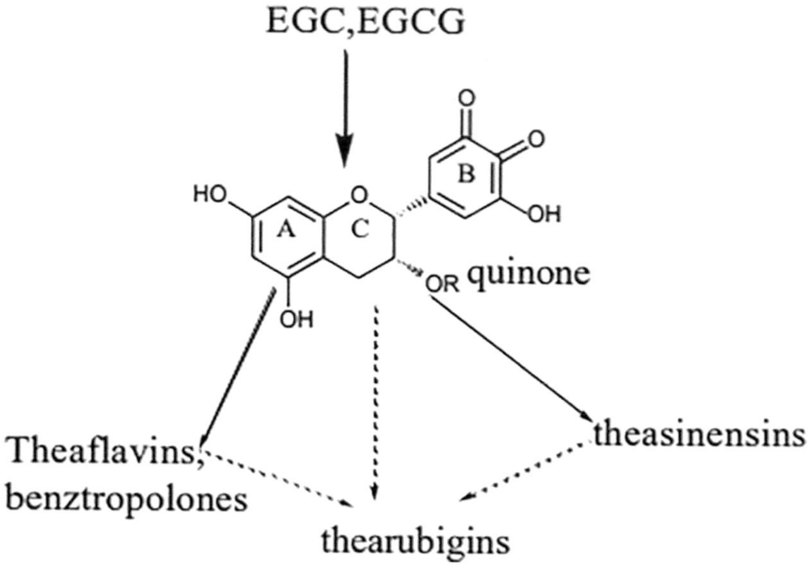

EGC,EGCG

Theaflavins,
benztropolones

quinone

theasinensins

thearubigins

Figure 16   Possible pathways to thearubigins, theaflavins and theasinensins (Haslam, 2003).

theaflavin, theaflavin gallate and thearubigins (Mathew and Parpia, 1971).

Theaflavins (TFs)

There are four main theaflavins (TFs) in black tea: Theaflavin TF1, theaflavin 3-gallate (TF2A), theaflavin 3′-gallate (TF2B) and theaflavin 3, 3′- gallate (TF3) (Figure 17) (Ostrowska, 2007).

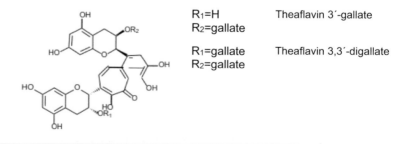

$R_1=H$	Theaflavin 3′-gallate
$R_2=gallate$	
$R_1=gallate$	Theaflavin 3,3′-digallate
$R_2=gallate$	

Figure 17   Structures of theaflavins (ChemSpider).

Thearubigins

Thearubigins consist of about 60% of the dry weight of black tea leaves with molecular weights of around 700–2,000 daltons (Haslam, 2003). They are poorly characterized chemically (Balentine, 1992). Theaflavin and theaflavin gallate were isolated by Roberts and Mayres from black tea, and further transformation of dicatechins and theaflavins resulted in the compounds that Roberts named thearubigins (Roberts, 1959). Figure 18 illustrates the possible intermediate structures for the formation of thearubigins (Haslam, 2003).

The acidic properties of thearubigins are related to the ionization of phenolic protons in the chromophore, for example -H* in structures A and B in Figure 12. Also, the acidity of thearubigins is behind the oxidative ring fission of phenolic nuclei in the chromophore (Figure 19).

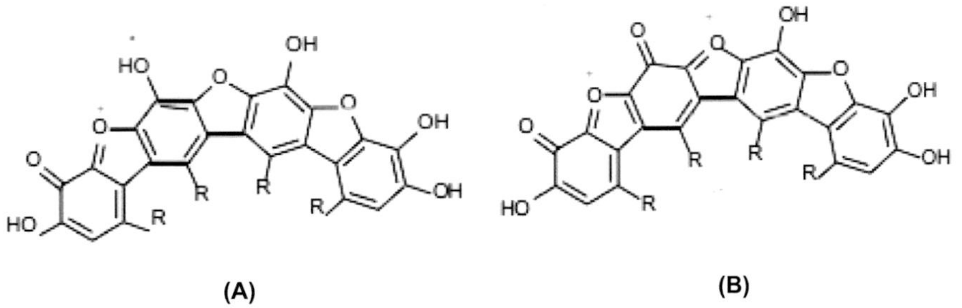

(A)   **(B)**

Figure 18   The possible intermediate structures (A, B) for the formation of thearubigins (Haslam, 2003).

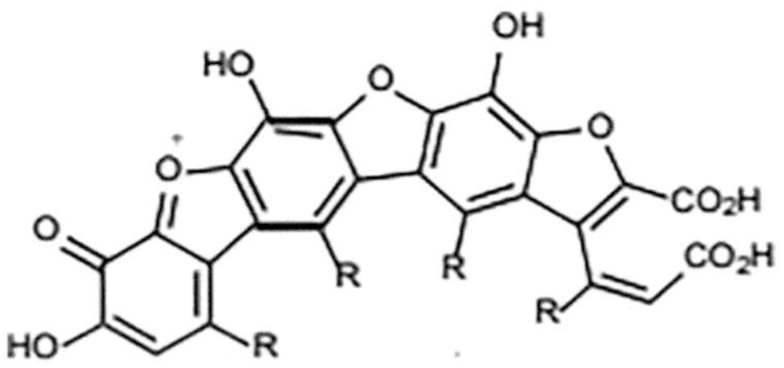

Figure 19   The structure explains the possible reason for the colour and acidity of thearubigins (Haslam, 2003).

Theasinensins

Another group of flavanols present in tea leaves are bisflava-
nols (Theasinensins). Theasinensins are including theasinensins A, theasinensins B, theasinensins C, theasinensins E and
oolongtheanin (Haslam, 2003) (Figures 20 and 21).

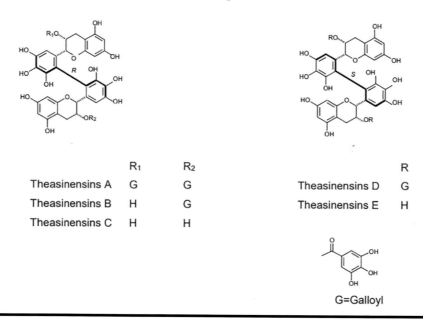

	R₁	R₂
Theasinensins A	G	G
Theasinensins B	H	G
Theasinensins C	H	H

	R
Theasinensins D	G
Theasinensins E	H

G=Galloyl

Figure 20   Structures of theasinensins A, B, C, D, E (Haslam, 2003).

Figure 21   Structure of oolongtheanin (ChemSpider).

Theasinensin A and D are two isomers, A has an R configuration and D has an S configuration and they are present in black tea. Theasinensin A with heating produce theasinensin D. Dehydrotheasinensin A under oxidation condition with phosphate buffer that keeps pH at 6.8 gives theasinensin A and theasinensin D. Oxidation-reduction of dehydrotheasinensin gives Theasinensin A and theasinensin D (Tanaka *et al.*, 2003).

Theaflagallins

Other black tea benzotropolen components are compounds called theaflagallins, synthesized as the result of mild oxidizing of gallocatechins and gallic acid mixtures.

Theaflagallins are including epitheaflagallin 3-*O*-gallate, epitheaflagallin and theaflagallin (Abudureheman *et al.*, 2022) (Figure 22).

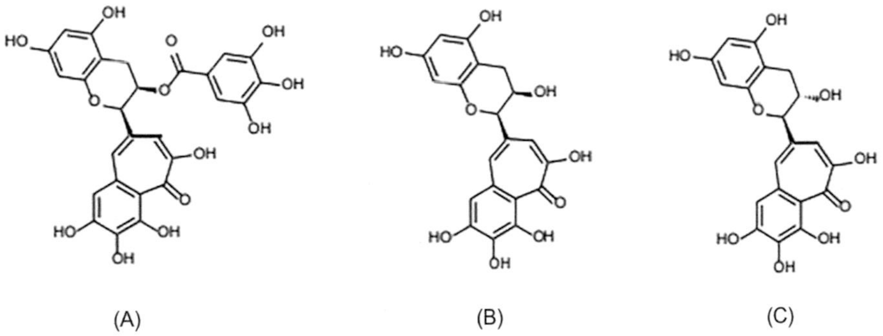

(A)   (B)   (C)

Figure 22   Structures of epitheaflagallin 3-O-gallate (A), epitheaflagallin (B) and theaflagallin (C) (Abudureheman *et al.*, 2022).

Galloyl-Glucose and Galloylquinic Acids (Hydrolysable Tannins)

Hydrolysable tannins, including 1, 4, 6-tri-*O*-galloyl-β-D-glucose, theogallin (a galloylquinic acid) and strictinin, were identified in black tea (Figure 23) (Hashimoto *et al.*, 2003).

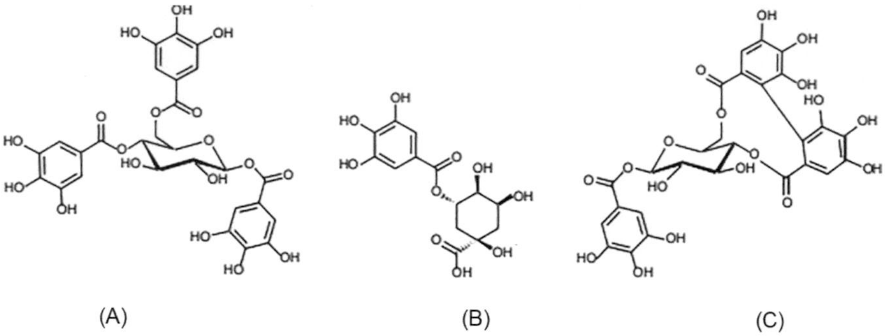

(A) (B) (C)

Figure 23 **Structures of 1, 4, 6-tri-O-galloyl-β-D-glucose (A), theogallin (B), strictinin (C) (Hashimoto et al., 2003).**

Chapter 3

Variability of Polyphenol Content in Foods

Polyphenols and phenolic compounds are a large part of the human diet, and therefore it is highly desirable to understand their biological functions and modes of activity (Perron, 2009; Ovaskainen *et al.*, 2008). Certain polyphenols such as quercetin are found in most fruits and vegetables, cereals, leguminous plants, tea, etc., while others are found in particular foods (flavones in citrus fruits, isoflavones in soya). Foods contain mixtures of polyphenols which are poorly characterized. Polyphenolic content in many plant products is unknown, and the data is often limited to one or a few varieties. Moreover, factors other than variety may affect the polyphenolic content of plants such as ripeness at the time of harvest, environmental factors, processing and storage conditions (Manach, 2004).

3.1 General Overview of a Typical Diet (How Much Do We Eat of What?)

As mentioned above, dietary phenolic acids and polyphenols are abundant in tea (black and green), fruits, vegetables, olive

DOI: 10.1201/9781003382652-3

Table 1   Phenolic Contents of Selected Beverages, Vegetables, Selected Fruits and Chocolate in Milligrams per Serving

Item	Phenolic Content (mg/ serving)	Item	Phenolic Content (mg/ serving)
Dark chocolate (60%) 40 g	951	Cranberries 55 g	373
Black tea 240 mL	943	Pear 166 g	317
Pomegranate juice 240 mL	616	Red grapes 80 g	296
Red wine 240 mL	431	Apple 138 g	256
Red onion 40 g	431	Cherries 73 g	231
Milk chocolate 40 g	394	Watermelon 286 g	183
Green tea 240 mL	247	Blueberries 70 g	181
Beet 85 g	201	Banana 118 g	163
Corn 91 g	129	Strawberries 122 g	162
Coffee 240 mL	115	White grapes 80 g	155
White wine 240 mL	92	Plum 66 g	149
Broccoli 71 g	75	Peach 117 g	82
Hot chocolate 240 mL	45	Orange 131 g	54
Carrot 71 g	33	Pineapple 78 g	52
Tomato 91 g	33	Lemon 85 g	24

Serving size is based on a typical beverage size (240 mL), piece of chocolate (40 g), or serving of vegetables.

oil, red and white wines as well as chocolate and the amount of them varies from a few micrograms to hundreds of milligrams or even grams per serving for the mentioned foods (Table 1), although the concentration of these compounds always varies for each type of food. People with diets rich in fruits and vegetables may consume about 5–10 g of polyphenols per day. The structural diversity of polyphenols makes the estimation of their content in food difficult (Scalbert and Williamson, 2000). The average content in some food servings is compared in the following charts (Table 1) (Perron, 2009).

3.2 Bioactivities of the Dietary Polyphenols

Epidemiologic studies have shown a correlation between increased consumption of phenolic and polyphenolic compounds and a reduced risk of cardiovascular disease (CVD), neurodegenerative diseases and certain types of cancers (Visioli and Galli, 2001; Andersen *et al.*, 2006; Singh *et al.*, 2008; Arts, 2008). The chemical structure of polyphenols affects their biological properties such as bioavailability, antioxidant activity, specific interactions with cell receptors and enzymes and other properties (Rice-Evans *et al.,* 1997; Tapiero *et al.*, 2002).

Polyphenols have been often evaluated for their biological activities in vitro on pure enzymes, cultured cells, or isolated tissues by using polyphenol aglycones or some glycosylated polyphenols available in dietary plants. Information about the biological properties of conjugated derivatives of polyphenols present in plasma or tissues is very scarce because of the lack of precise identification and availability of commercial standards (Manach *et al.*, 2004; Liu and Hu, 2007; Zumbé, 1998; Yang *et al.*, 2008)

3.2.1 Oxidative Stress

Reactive oxygen species (ROS) such as hydroxyl radical (OH·), hydrogen peroxide (H_2O_2) and superoxide anion radical (O_2^-) are responsible for oxidative damage in cells such as damaging proteins, lipids and DNA (Wiseman *et al.*, 1996; Bennett, 2001; Darley-Usmar and Halliwell, 1996; Hensley *et al.*, 2000). The free radical theory of aging was proposed by Denham Harman more than 50 years ago and the phenomenon has been termed "oxidative stress" by Helmut Sies in 1985 (Sies, 1997). Oxidative damage of DNA is a cause of cancer, aging, neurodegenerative diseases such as Alzheimer's and Parkinson, and CVDs such as arteriosclerosis (Bennett, 2001; Darley-Usmar and Halliwell, 1996; Hensley *et al.*, 2000; Ofodile, 2006; Brewer, 2010). Prevention of oxidative stress caused by ROS is important for the prevention and treatment of such diseases (Behl, 2005; Prasad *et al.*, 1999; Boldogh and Kruze, 2008; Polidori, 2004).

3.2.2 Antioxidant and Free Radical Scavenging Properties of Polyphenols

Antioxidants are reducing agents that are capable of delaying or inhibiting the oxidation of other molecules. In oxidation reactions, free radicals can be produced, which start chain reactions that damage cells (Apak *et al.*, 2007; Javanmardi *et al.*, 2003). Antioxidants such as polyphenols terminate these chain reactions through their radical scavenging property and inhibit other oxidation reactions by being oxidized themselves (e.g., oxidation by H_2O_2, Fe^{3+} and Cu^{2+}) (Apak *et al.*, 2007; Patel, 2010; Chanwitheesuk *et al.*, 2005). Cells and tissues are threatened by the damage caused by free radicals which are produced during metabolism or induced by exogenous damage. One of the most important reasons why free radicals interfere with cellular functions seems to be due to lipid peroxidation which causes cellular membrane damage leading to a shift in the net charge of the cell, thus changing the osmotic pressure within the cell, resulting in swelling and cell death (Nijveldt *et al.*, 2001; Biswas *et al.*, 2005; Grijalba *et al.*, 1998; Sánchez-Moreno *et al.*, 1999).

3.2.3 *The Ways Polyphenols Exhibit Their Antioxidant Activity*

Flavonoids and other polyphenols exhibit their antioxidant activity in several ways:

1. Radical scavenging activity toward reactive species (e.g., ROS) or lipid peroxidizing radicals such as R·, RO· and ROO· Radical scavenging action generally occurs through hydrogen atom transfer or electron donation (Figure 24).
2. Prevention of the transition metal–catalyzed production of reactive species.
3. Interaction with other antioxidants (such as cooperative actions), localization and mobility of the antioxidant in the microenvironment (Apak *et al.*, 2007; Martinez-Gomez *et al.*, 2020).

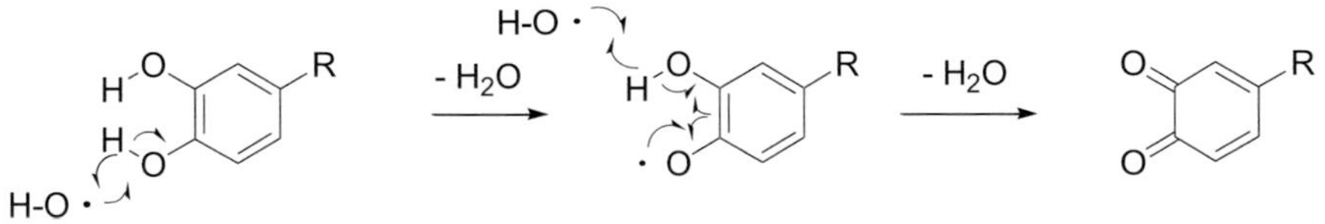

Figure 24   Radical scavenging mechanism of polyphenols (Martinez-Gomez et al., 2020).

Polyphenols exhibit a wide range of biological effects as a consequence of their antioxidant properties (Urquiaga and Leighton, 2000; Su *et al.*, 2007; Aiyegoro, and Okoh, 2009). The radical scavenging properties of polyphenols are important for preventing diseases associated with oxidative damage of membranes, proteins, lipids and DNA (Ferguson, 2001). As mentioned earlier, studies suggest that consumption of dietary polyphenols is associated with reduced risk of various cancers, prevention of neurodegenerative and heart diseases as well as prevention and minimizing the menopausal symptoms (Hodek *et al.*, 2002; Zern *et al.*, 2005; Williamson and Manach, 2005).

3.2.4 *Potential Mode of Action of Phenolic Compounds*

Besides the strong antioxidant capacities, tea polyphenols have been largely studied for other properties by which cell activities are regulated (Han *et al.*, 2007; Masella *et al.*, 2005). They might interfere through several mechanisms that lead to the development of tumours, inactivating carcinogens, inhibiting the expression of mutant genes and enzymes involved in the activation of procarcinogens (Urquiaga and Leighton, 2000; Barbosa, 2007; Kampa *et al.*, 2007). Studies have shown that polyphenols, particularly flavonoids, inhibit the initiation, promotion and progression of tumours, possibly by a different mechanism other than their antioxidant protective effects on DNA and gene expression (Urquiaga and Leighton, 2000; Lin *et al.*, 1999; Lee *et al.*, 2006; Yang *et al.*, 1998).

Flavonoids and their metabolites exert modulatory actions through the activation of different protein kinases such as mitogen-activated protein kinase (MAP kinase) (Williams *et al.*, 2004). There are several types of MAP kinase such as extracellular signal-regulated kinase (ERK), c-Jun NH2-terminal kinase (JNK) and P38, and each of them is responsible for different functions for example apoptosis. These MAP kinases phosphorylate different transcription factors such as NF-κB and activator protein-1(AP-1). For example, the NFκB regulates the expression of cytokines, growth factors and inhibitors of apoptosis. MAP kinase is a protein kinase which can be activated by different mitogens (signals that come to the cell). Once MAP kinase is activated then it changes the spectrum of transcription in a cell as a response to stimuli (Crozier *et al.*, 2009; Schroeter *et al.*, 2002; Spencer *et al.*, 2003).

There are many examples of polyphenols that play role in the inhibition or activation of enzymes by which cell activities are regulated. The research for understanding the mechanism of actions observed for health benefits has recently intensified. As an illustration, the inhibitory effect of (-)-epigallocatechin-3-gallate on Ultraviolet Bactivated phosphatidylinositol 3-kinase (PI3K) in mouse epidermal JB6 Cl 41 cells was shown in a study. The researchers suggest that because PI3K is an important factor in carcinogenesis, the inhibitory effect of these polyphenols on the activation of PI3K and its downstream effects may further explain the anti-tumour promotion action of the tea polyphenols (Nomura *et al.*, 2001). In another study, results provide the first evidence that 5-caffeoylquinic acid could protect against environmental carcinogen-induced carcinogenesis and suggest that the chemopreventive effects of 5-caffeoylquinic acid may be through its upregulation of cellular antioxidant enzymes and suppression of ROS-mediated NF-κB, AP1 and MAPK activation (Feng *et al.*, 2005). Various examples of the biological activities of dietary polyphenols and their health effects are illustrated in the following (Figure 25) (Han *et al.*, 2007).

3.2.5 *Biological Activities of Flavonoids*

Flavonoids are considered the major bioactive components of many medicinal plants. Besides the antioxidant activity, they show biological activities and clinical effects such as anti-atherosclerotic effects, anti-tumour, anti-mutagenic, anti-thrombotic, anti-inflammatory, anti-osteoporotic, anti-allergic and anti-viral activity (Fabre *et al.*, 2001; March *et al.*, 2006; Le Marchand, 2002; Hollman and Katan, 1999; Merken and Beecher, 2000). For example, hesperetin was found to inhibit the replication of herpes simplex virus type 1,

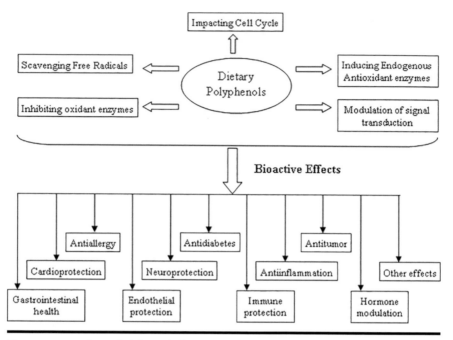

Figure 25   Bioactivities of dietary polyphenols (Han *et al.*, 2007).

poliovirus type 1 and parainfluenzavirus type 3. Hesperidin (hesperetin-7-O-rutinoside) and naringin have inhibitory activity on rotavirus infection (Paredes *et al.*, 2003; Garg *et al.*, 2001).

3.2.6 Biological Activities of Isoflavones

Four phenolic compounds classified as phytoestrogens that possess estrogen-like biological activity are isoflavones, stilbenes, coumestans and lignans. They mimic estradiol-like effects in several tissue/tissues of the mammalian body. Isoflavones are particularly abundant in plants belonging to the Fabaceae family, especially in soybeans. Due to the structural similarity of these polyphenolic compounds to mammalian ß-estradiol (Figure 26), isoflavones are known to possess estrogenic activities. They have several health benefits such as anticarcinogenic activity, protection against CVDs, and protection

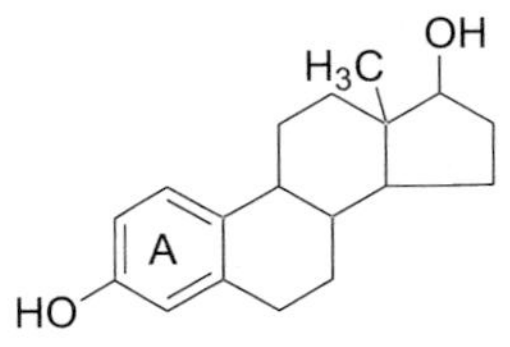

Figure 26   17ß-estradiol (Vincent and Fitzpatrick, 2000).

against osteoporosis and menopausal symptoms which are due to their estrogenic action (Cavaliere *et al.*, 2007; Vincent and Fitzpatrick, 2000).

3.2.7 Biological Activities of Anthocyanidins

Anthocyanidins may enhance anti-inflammatory and neu-roprotective effects. They can inhibit lipid peroxidation and enzymes associated with tumour formation and reduce the risk of coronary heart diseases (Mazza, 2007; Montoro *et al.*, 2006).

3.2.8 Biological Activities of Tannins

Tannins are reported to possess biologically multifunctional properties including antioxidant, anti-bacterial, anti-viral, anti-inflammatory, anti-tumour activities and inhibitory effects on various enzymes (Soong and Barlow, 2005).

3.3 Biological Activities of Tea Polyphenols

Tea is one of the healthiest beverages in the world. In tradi-tional treatment, green tea is widely used for its health effects against many diseases. Studies have shown that the health effects of tea are related to its polyphenol contents and the properties of these compounds. They also chelate heavy met-als and protect the body against the toxicity of heavy metals (Kuroda and Hara, 2004).

3.3.1 Anti-Atherosclerosis and Cardioprotection Effects

Epidemiological, clinical and experimental studies have established a positive correlation between green tea consumption and cardiovascular health. Catechins, the major polyphenolic compounds in green tea, exert vascular protective effects through multiple mechanisms, including antioxidative, anti-hypertensive, anti-inflammatory, anti-proliferative, anti-thrombogenic and lipid-lowering effects (Velayutham *et al.*, 2008). Another example of tea polyphenols is quercetin which inhibits lipid peroxidation determining a qualitative reduction of oxidized low-density lipoproteins (oxy-LDL) (Han *et al.*, 2007).

3.3.2 Anti-Carcinogenic Effects

Tea flavonoids and especially catechins play a role in protecting the body against chemicals that induce cancers (Liao *et al.*, 2001; Dufresne and Farnworth, 2001; Katiyar and Mukhtar, 1996; Yang and Wang, 1993; Mukhtar and Ahmad, 1992; Ahmad *et al.*, 1996). They are radical scavengers and directly neutralize procarcinogens by their ROS scavenging action before cell damage occurs (Yoshioka *et al.*, 1996; Yamada and Tonita, 1994) thus reducing the risk of cancer (Blot *et al.*, 1996). Some studies show that epigallocatechin gallate is more effective against cancer than the other polyphenols of green tea (Katiyar and Mukhtar., 1996; Yang and Wang, 1993; Mukhtar and Ahmad, 1992; Ahmad *et al.*, 1996). It also induces apoptosis and inhibits cell growth (Yang *et al.*, 1998; Ahmad *et al.*, 1997; Zhang *et al.*, 2000). Green tea polyphenols help to protect against various cancers through different and multiple mechanisms aforementioned in Figure 8. Illustrations of health benefits of tea polyphenols against cancer diseases are lung cancer (Okabe *et al.*, 1997), breast cancer (Nakachi *et al.*, 1998), skin cancer (Chung *et al.*, 1999), stomach cancer (Oguni *et al.*, 1989; Hoshiyama *et al.*, 2002) and prostate cancer (Klein and Fischer, 2002; Paschka *et al.*, 1998).

3.3.3 Neuroprotective Effects

There has been considerable interest in the neuroprotective effects of some tea polyphenols and the mechanism of action of these effects. Such polyphenols had been considered as therapeutic agents for altering brain aging processes and as possible neuroprotective agents in progressive neurodegenerative disorders such as Parkinson's and Alzheimer's diseases (Han *et al.*, 2007). Some tea polyphenols which show neuroprotective effects are epigallocatechin gallate and epicatechin gallate (Levites *et al.*, 2002), catechin, quercetin, (Mercer *et al.*, 2005), epicatechin and kaempferol (Schroeter *et al.*, 2001).

3.3.4 Anti-Allergic, Anti-Inflammatory and Immuno-Modulatory Effects

Decades of research on polyphenols have led to several insights regarding the effects of polyphenols on immune function (Ding *et al.*, 2018). Each type of polyphenol targets and binds to one or more receptors on immune cells and thus triggers intracellular signaling pathways that ultimately regulate the host immune response. Dietary interventions that involve polyphenols may modulate immune responses (Dias *et al.*, 2018).

Tea contains a various range of polyphenols with anti-allergic, anti-inflammatory, and immunomodulatory properties including quercetin (Winiarska-Mieczan, 2018; Bharadwaz and Bhattacharjee, 2012), tannic acid, green tea catechins epigallocatechin-3-gallate (Thichanpiang and Wongprasert, 2015; Samarghandian *et al.*, 2017), theaflavins (Gothandam *et al.*, 2019) and thearubigins (Winiarska-Mieczan, 2018; Bharadwaz and Bhattacharjee, 2012) present in black tea as well as quercetin (Greca and Zarrelli, 2012; Winiarska-Mieczan, 2018; Bharadwaz and Bhattacharjee, 2012).

3.3.5 Gastrointestinal Protection Effects

The effect of green tea drinking on reducing human cancer risk is unclear, though a protective effect has been reported in numerous animal studies and several epidemiologic investigations (Gothandam *et al.*, 2019). Accumulating evidence shows that intake of green tea rich in epigallocatechin gallate showed a protective effect in various human gastric, epithelial and colon cells (Yang *et al.*, 2016; Kim *et al.*, 2004).

Catechins have been a major focus of both basic and clinical research in recent years, due to the high level of ingestion, particularly in Asian societies. With a cohort of tea-drinking Chinese individuals, Ji *et al.* (1997) performed a large population-based case–control study on the effects of green tea consumption on pancreatic enzymes and colorectal cancer (CRC). After adjusting for age, income, education and cigarette usage, green tea consumption was inversely associated with cancer incidence, particularly pancreatic cancer and CRC. Consumption of green tea by men and women at the highest levels resulted in a significantly lower odds ratio (OR) than in controls (colon cancer 0.82 for males, 0.67 for females, rectal cancer 0.72 for males, 0.57 for females, and pancreatic cancer 0.63 for males, 0.53 for females) (Gerald *et al.*, 2006).

3.3.6 Modulation of Hormonal Effects

Tea contains phytoestrogens. These estradiol-like molecules affect several mammalian tissues. The potential health benefits of these molecules are due to their ability to mimic estrogenic actions. Tea contains up to 20 µg/100 g phytoestrogens mainly lignans (Gunter *et al.*, 2008; Mostrom and Evans, 2011).

As described in Section 3.2.6, phytoestrogens are estrogen-like polyphenolic compounds produced naturally by certain plants and have structural similarity to the mammalian steroid hormone 17 β-estradiol (Figure 26). Four phenolic compounds classified as phytoestrogens are isoflavones, stilbenes, coumestans and lignans (Figure 27) (Speirs, 2008; Fatima *et al.*, 2018). These compounds possess several pharmacological applications including cardioprotection, antimicrobial, anticancer, anti-obesity, antiosteoporosis, antidiabetic, and neuroprotection which are due to their estrogenic action (Fatima *et al.*, 2018; Wang *et al.*, 2021).

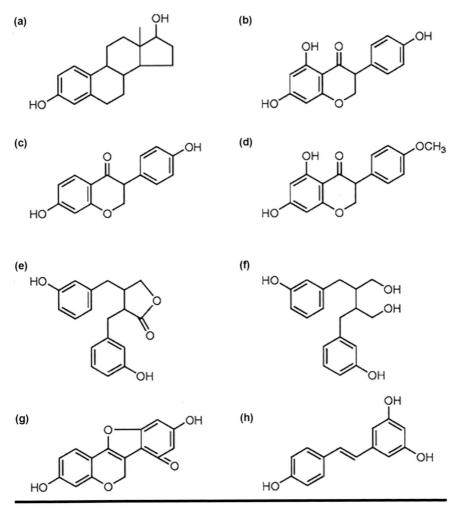

Figure 27   Structural comparison of 17β-estradiol (a) with the principal classes of phytoestrogen; the isoflavones genistein (b), daidzein (c), and biochanin A (d); the lignans enterolactone (e) and enterodiol (f); the coumestan coumestrol (g); and the stilbene resveratrol (h) (Speirs, 2008).

3.3.7 Cardiovascular Protection Effect

CVDs are public health issues with high mortality, taking an estimated 17.9 million lives each year (WHO). Epidemiological studies have revealed that regular tea drinking is inversely associated with the risk of CVDs, especially in habitual tea drinkers, and have suggested a protective role of tea and its bioactive components in cardiovascular health (Lange, 2022). Flavonoids and in particular flavan-3-ols, found in both green and black tea, have been suggested to play a primary role in the reduction of CVDs risk (Lange, 2022). In vitro and in vivo experimental studies have shown that tea and its bioactive compounds are effective in protecting against CVDs. The relevant mechanisms include reducing blood lipids, alleviating ischemia/reperfusion injury, inhibiting oxidative stress, enhancing endothelial function, attenuating inflammation, and protecting cardiomyocyte function (Cao *et al.*, 2019).

An intake of tea at moderate levels (around 2–3 cups per day) seems to have the potential to decrease CVD risk and progression due to its flavonoid content. Because of the widespread consumption of tea and the high global prevalence of hypertension, the beneficial effects of tea may nevertheless be important for cardiovascular health at the population level. Even small beneficial effects in humans may shift the population distribution of CVDs risk, with major implications for public health (Keller and Wallace, 2021; Cao, 2019).

3.3.8 Tea for the Prevention and Management of Diabetes Mellitus and Diabetic Complications

Diabetes mellitus is a group of metabolic diseases characterized by hyperglycemia resulting from defects in insulin secretion, insulin action, or both. The chronic hyperglycemia of diabetes is associated with long-term damage, dysfunction, and failure of various organs, especially the eyes (diabetic retinopathy), kidneys (diabetic nephropathy), nerves (diabetic neuropathy), liver (diabetic hepatopathy), heart and blood vessels (diabetic CVD) (American Diabetes Association, 2009).

Diabetes occurs when the pancreas, a gland behind the stomach, does not produce enough insulin or the body cannot use insulin. Several reasons are involved in the development of diabetes, for example, autoimmune destruction of the β-cells of the pancreas usually leads to absolute insulin deficiency. In type 2 diabetes (T2DM), diminished tissue responses to insulin at one or more points in the pathways of hormone action result in resistance to insulin action. The deficient action of insulin on target tissues results in irregularities in carbohydrate, fat, and protein metabolism (American Diabetes Association, 2009).

The effects of tea on diabetes mellitus and its complications have been widely investigated. Epidemiological studies have suggested that tea consumption (black tea, green tea, and oolong tea) has an inverse relation with the risk of diabetes mellitus and its complications (Meng *et al.*, 2019, Jing *et al.*, 2009, Vasto *et al.*, 2014, Greca and Zarrelli, 2012). Epidemiological, experimental, and clinical studies reveal that the protective effects of tea against diabetes mellitus and its complications happen via several possible mechanisms (Mieczan *et al.*, 2021). These include enhancing insulin action, improving insulin resistance, activating the insulin signaling pathway, playing an insulin-like role, protecting islet β-cells, improving oxidative stress by scavenging free radicals, and decreasing inflammatory response (Figure 28) (Meng *et al.*, 2019). The molecular mechanisms of EGCG on diabetes mellitus and its complications are summarized in Figure 29 (Meng *et al.*, 2019). Besides, tea has synergistic effects with certain antidiabetic drugs. Clinical trials also confirmed that tea has protective effects on patients with diabetes mellitus and diabetic complications (Meng *et al.*, 2019). Since different types of tea have different bioactive compounds, therefore tea could be developed into nutraceuticals for the prevention and management of diabetes mellitus and diabetic complications. Clinical trials should be carried out to verify the protective doses of tea on diabetes mellitus and its complications in humans. In addition, special attention should be paid to the safety of tea and tea products.

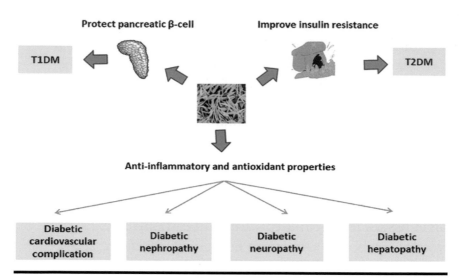

Figure 28 The association between tea and diabetes and its complications. Tea has effects on type 1 diabetes mellitus (T1DM) and type 2 diabetes mellitus (T2DM) by protecting pancreatic β-cells and ameliorating insulin resistance (Meng et al., 2019).

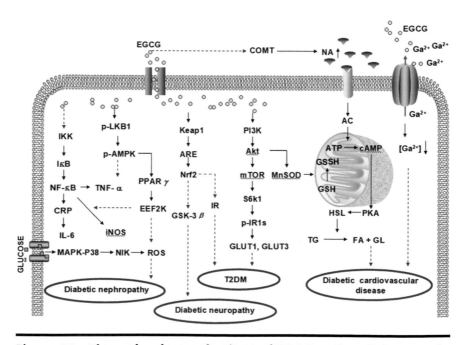

Figure 29 The molecular mechanisms of EGCG against diabetes mellitus and its complications.

EGCG has shown effects against T2DM by improving IR, against diabetic CVD by decreasing TG and [Ga2+], against diabetic nephropathy by decreasing ROS, and against diabetic neuropathy by increasing Nrf2. The arrow means the direction of actions, the black full lines indicate upregulation and the red dotted lines refer to downregulation or inhibition. CRP, C-reactive protein; MAPK p38-NIK, NF-κB inducing kinase; LKB1, kelch-like ECH-associated protein-1; EEF2K, eukaryotic elongation factor-2 kinase; ARE, antioxidant-responsive element; GSK-3β, glycogen synthase kinase-3β; IR, insulin resistance; MnSOD, Mn superoxide dismutase; NA, noradrenalin; s6k1, ribosomal protein S6 kinase 1; AC, adenylate cyclase; HSL, hormone-sensitive lipase; TG, triglyceride; FA, fatty acid; GL, glycerinum; GSH, glutathione; GSSH, oxidized glutathione; mTOR, the target of rapamycin; EGCG, epigallocatechin gallate; IKK, IκB kinase; NF-κB, nuclear factor-κB; iNOS, inducible nitric oxide synthase; TNF-α, tumor necrosis factor-α; Nrf2, nuclear factor-erythrocyte-associated factor 2; PI3K, phosphatidylinositol 3-hydroxykinase; Akt, protein kinase B; AMPK, adenylic acid-activated protein kinase; T2DM, type 2 diabetes mellitus; GLUT, glucose transporter type; PKA, protein kinase A; ATP, adenosine triphosphate; cAMP, cyclic adenosine monophosphate; COMT, catechol-O-methyltransferase, an enzyme responsible for the degradation of noradrenaline (Meng et al., 2019).

3.3.9 Tea and Osteoporosis

Osteoporosis is a kind of disease where the density of bones is reduced leading to a risk of fracturing. This disease is particularly observed in elderly women and is caused by a lack of estrogen. Recent studies have shown that regular drinking of tea can protect against osteoporosis in postmenopausal women (Kanis *et al.*, 1999). Bad living habits such as smoking and drinking alcohol may reduce the effect of tea consumption on osteoporosis (Sun *et al.*, 2017).

Experimental studies have shown that consuming one to six cups of tea per day may reduce the risk of bone fracture by increasing the level of minerals in bones (Hegarty *et al.*, 2000; Hammett-Stabler, 2004). This could be through the function of isoflavonoids contained in tea with the highest estrogenic properties such as genistein and daidzein. The exact mechanism of the relationship between tea consumption and osteoporosis still needs further research (Sun *et al.*, 2011).

3.3.10 The Anti-Obesity Effects of Tea

Tea can help in preventing and combatting obesity. The World's leading cause of death are diseases related to metabolic dysfunctions. In recent years, obesity has become an extremely widespread phenomenon which is still poorly understood from an etiologic-mechanistic perspective. In 2016 it was estimated that 1.9 billion adult people around the world were overweight, of whom 650 million were obese (Haider and Larose, 2019). Effective pharmacological solutions to prohibit the development of obesity are very few. Obesity is a major risk of life-threatening diseases and can facilitate type 2 diabetes, cause inflammations, cardiovascular and reproductive diseases, non-alcoholic fatty liver disease (NAFLD), and cancer (Skrypnik *et al.*, 2017, Sung *et al.*, 2018). In addition to health issues, obesity causes problems in mobility, and social relationships and affects the overall quality of life (Mohamed *et al.*, 2014).

Natural products have been the basis of oriental medicine for the treatment of diseases. Tea could be a promising candidate for the prevention and treatment of several diseases including obesity (Sirotkin and Kolesarova, 2021).

The anti-obesity potential of green tea catechins, particularly EGCG, has been shown in cell culture, animal and human studies (Wang *et al.*, 2014). An experimental study shows that the daily intake of EGCG can reduce body weight by approximately 20%– 30% within 2–7 days (Kao *et al.*, 2000). Tea consumption can reduce fat storage through the suppression of adipocyte functions, and support of gut micro-biota. In addition, tea can prevent obesity via the reduction of appetite, food consumption and food absorption in the gastrointestinal system and through changes in fat metabolism (Sirotkin and Kolesarova, 2021).

Tea of all types has a positive effect on health and weight loss. In medication, tea can be substituted by its functional components EGCG and caffeine. It is necessary to remember that drinking about four cups of strong tea a day (1–2 g of tea containing 100–600 mg polyphenols) for a minimum of 8 weeks can result in weight loss (Sirotkin and Kolesarova, 2021).

3.3.11 Skin Protection Effects of Tea against Aging and Cancer

Studies indicate that exposure of the skin to environmental factors such as solar ultraviolet (UV) irradiation, air pollutants or smoking can generate ROS which causes oxidative stress and aging and induces direct and indirect biological harmful effects (Figure 30) (Lee *et al.*, 2022). These effects lead to skin diseases including inflammation, DNA damage, skin aging, dysregulation of cellular signaling pathways and immunosuppression thereby resulting in melanoma and non-melanoma skin cancer. The regular intake of polyphenols, which are widely present in green tea such as EGCG and (–)-catechin, could act as protective agents against the adverse effects of UV irradiation and improve skin aging (Afaq and Katiyar, 2011, Lee *et al.*, 2022). Polyphenols are antioxidant molecules with anti-inflammatory, and antineoplastic properties. Studies suggest that polyphenols could be used for the prevention of sunburns as polyphenols decrease the damaging effects of ultraviolet A (UVA) and ultraviolet B (UVB) radiation on the skin. Based on the evidence, polyphenols in both oral and topical forms protect from UV damage and sunburn and thus are beneficial to skin health (Saric and Sivamani, 2016). The mechanism of action of certain polyphenols including EGCG and (–)-catechin in improving skin aging has been illustrated in several studies (Lee *et al.*, 2022). The new information on the mechanisms of action of the polyphenols supports their potential use in skin photoprotection and prevention of photo-carcinogenesis in humans (Afaq and Katiyar, 2011). Regardless of the findings, strategies such as prevention from sun exposure, and the use of sun-protective clothing and sunscreens are important first-line methods for sun protection (Saric and Sivamani, 2016).

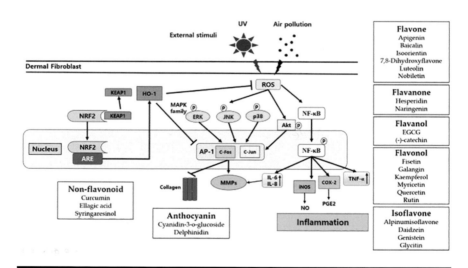

Figure 30   Diagram of several molecular mechanisms in the skin dermis exposed to external stimuli. External stimuli such as UV radiation or air pollutants can cause direct damage to the DNA and produce ROS. These can further stimulate many inflammatory responses and the MAPK family, which can lead to photoaging through inflammation and collagen degradation. The KEAP1-NRF2 stress response pathway is the principal inducible defence against oxidative stresses. Under homeostatic conditions, KEAP1 regulates the activity of NRF2. In response to stress, an intricate molecular mechanism facilitated by sensor cysteines within KEAP1 allows NRF2 to escape ubiquitination, accumulate within the cell, and translocate to the nucleus, where it can promote its antioxidant transcription program (Lee *et al.*, 2022).

Chapter 4

Antiviral Effects of Tea

Several plants derived virucidal and antiviral compounds have been identified. The potential application of these compounds for the treatment of viral diseases has been suggested. Phenolics, carotenoids, terpenoids, and alkaloids are well-known antiviral phytochemicals. The virucidal and antiviral activities of naturally derived phenolic acids, flavonoids, stilbenes, coumarins, and tannins are known. They have been reported to prevent viral absorption and entry to host cells and inhibit viral genome and protein synthesis in infected cells (Takeda *et al.*, 2021). Black and green teas are widely available and are relatively cheap to purchase in both developed and developing countries and possess a broad range of antiviral spectrum on both enveloped and non-enveloped viruses (Keflie and Biesalski, 2021).

Studies have confirmed the antiviral activities of tea polyphenols against various viruses especially positive-sense single-stranded RNA viruses (Singh *et al.*, 2021). Some of those include Porcine Reproductive and Respiratory Syndrome Virus (PRRSV) (Wang *et al.*, 2021), Influenza A Virus and B Virus (Yang *et al.*, 2014), Herpes Simplex Virus (HSV) (Ishimoto *et al.*, 2021; Pantel *et al.*, 2018), Human Immunodeficiency Virus (HIV) (Haubera *et al.*, 2009), West Nile Virus (WNV) (Vázquez-Calvo *et al.*, 2017),

Zika Virus (ZIKV) (Vázquez-Calvo *et al.*, 2017), Dengue Virus (DENV) (Vázquez-Calvo *et al.*, 2017), Hepatitis B Virus and Hepatitis C Virus (HBV and HCV) (Ghosh *et al.*, 2020; Ciesek *et al.*, 2011). Chikungunya Virus (CHIKV) (Ghosh *et al.*, 2020), Bovine Coronavirus, Bovine Rotavirus (Ishimoto *et al.*, 2021). SARS-CoV-2 (Mhatre *et al.*, 2021; Bosch *et al.*, 2003) and Covid19 (Keflie and Biesalski, 2021).

4.1 Antiviral Properties of Epigallocatechin-3-gallate (EGCG)

Several studies have been done for identifying the antiviral activity of tea polyphenols, especially EGCG. It was shown that EGCG can inhibit PRRSV infection regardless of whether it was administered before or after infection; a concentration of $125\,\mu$M was enough to completely inhibit the infectivity of viral cells (Mhatre *et al.*, 2021).

EGCG was found to be able to inhibit the Hepatitis C Virus (HCV) infection by attaching to the target cell and preventing the spread of the infection to different cells (Mhatre *et al.*, 2021).

EGCG inhibits the entry of the ZIKV by interacting with the lipid bilayer. An in vivo study showed that EGCG can cross the placental barrier and spread to the brain, eyes and heart of the fetus. This is a very important point for considering its administration because it is effective in pregnant women and possibly the fetus too (Mhatre *et al.*, 2021).

EGCG inhibits both the intracellular CHIKV replication and the extracellular infection in the pre and post-stages of viral infection (Mhatre *et al.*, 2021).

EGCG inhibits the intermediate stages of the influenza virus cycle by inhibiting hemagglutination activity (Mhatre *et al.*, 2021).

EGCG is effective against HIV-1 and inhibits viral replication by acting at various stages. It blocks the interaction of gp120 with CD4 by interfering with reverse transcriptase. A report suggested that EGCG inhibits the production of p24 antigen on isolated CD4 receptor cells, macrophages and CD4±T cells depending on its dose. In another study, it was shown that even at concentrations obtained by the consumption of green tea, EGCG was effective in inhibiting gp120-CD4 attachment (Mhatre *et al.*, 2021).

Table 2 summarizes the known antiviral activities of EGCG (Mhatre *et al.*, 2021). EGCG is an antiviral agent with a different mechanism for various viruses. The reason behind the

Table 2   The Summary of Known Antiviral Activities of EGCG

Virus	Viral Genome	Experimental Antiviral Effects of EGCG
PRRSV	ssRNA	EGCG effectively inhibits PRRSV infection. It prevents MARC-145 cells from getting infected by PRRSV
HCV	ssRNA	EGCG prevents infection by inhibiting the entry of HCV into hepatoma cell lines and primary human hepatocytes
HIV	ssRNA	EGCG interferes with the interaction of host cell receptors and virus envelope and inhibits the entry of the virus into target cells
Zika virus	ssRNA	Cells pre-treated with EGCG showed no virus infection
Chikungunya (CHIKV)	ssRNA	The entry, replication, and release of CHIKV are inhibited in vitro by EGCG
West Nile Virus (WNV)	ssRNA	EGCG has a direct effect on WNV when treated at the early stages of the infection
Dengue	ssRNA	EGCG interacts with the virus molecule causing virus deformation
Influenza A/H1N1, A/H3N2, B	ssRNA	EGCG inhibits acidification of intracellular endosome compartments essential in the fusion of membranes of virus and host cell
Ebola	ssRNA	EGCG inhibits the HSPS5 protein of the host which is the target of Ebola virus treatments thus reducing the virus multiplication once infected
Murine norovirus and feline calicivirus	ssRNA	EGCG at 100 IAA was found to be the most potent prophylactic agent when compared with other flavonoids

difference in the antiviral activity is the number of hydroxyl groups present on the benzene ring and galloyl group, together with the pyrogallol group which is responsible for exhibiting diverse mechanisms (Mhatre *et al.*, 2021).

4.2 Antiviral Properties of Theaflavins (TFs)

TFs (Figure 17) have antiviral activity against the viral particles of both enveloped and non-enveloped virus species, including influenza virus, HIV, rotavirus, enterovirus, calicivirus, and Hepatitis C Virus (Zhao *et al.*, 2021; Mhatre *et al.*, 2021). Table 3 gives a summary of a few reports on the antiviral activities of TFs (Mhatre *et al.*, 2021). Antiviral properties of derivatives of theaflavin polyphenols have also been explored in several viral diseases (Mhatre *et al.*, 2021).

TF1, TF2, and TF3 inhibit the HSV viral lytic cycle strongly with TF3 being the most potent antiviral compound (Mhatre *et al.*, 2021). Experiments have confirmed the prophylactic antiviral activity of TF3. There is more than 99% inhibition of the infectivity of viral molecules when they are treated with TF3 at 50 μM for 1 hour, confirming that TF3 is a potent antiviral compound (Mhatre *et al.*, 2021).

TFs demonstrate antiviral activity against the influenza virus via neuraminidase inhibition and prevention of the virus from adsorbing into the MDCK cells hence inhibiting the hemagglutinin of the virus. TF3 is more potent than other derivatives and almost comparable to the control oseltamivir carboxylate (Mhatre *et al.*, 2021).

TF2B is the most potent among all the other tea polyphenols in the inhibition of HIV at a concentration of 1 μM and a selectivity index greater than 200. The number of galloyl groups on the TF has a direct relation to its activity and it is estimated that these molecules interact with the gp41 six-helix bundle to prevent viral entry into the host cell. At a relatively higher concentration, TFs inhibit reverse transcriptase in HIV (Liu *et al.*, 2005).

TFs showed to have the best antiviral activity among 2,080 small molecules tested against caliciviruses. The hydroxyl groups in TFs are more essential in exhibiting the activity than the galloyl groups (Mhatre *et al.*, 2021).

TFs are active against HCV viral particles and inhibit their binding to the surface of the receptor, with TF3 being a better anti-HCV active compound than TF1 and TF2 (Ghosh *et al.*, 2020, Mhatre *et al.*, 2021; Ishimoto *et al.*, 2021).

Table 3   Summary of Known Antiviral Activity of TFs

Virus	Genome	Experimental Antiviral Effects of TFs
Sindbis Virus	ssRNA	TFs extract inhibits the viral infection by 99% at a concentration of 14.6 mM
TMV	ssRNA	TFs interfere with the viral replication cycle by attachment to TMV–RNA complex
Influenza A and B	ssRNA	TF1 has an IC50 value of 16.21 pgimL against the virus, which is the best among 13 flavonoids studied using cytopathic effect inhibition assay
HSV	ssRNA	TF3 in combination with acyclovir shows an increased 21.8% inhibition than acyclovir alone in WST-1 assay
Rotavirus and coronavirus	ssRNA	Synergistic activity when all IFS were administered in vitro
HCV	ssRNA	TFs act directly on the viral particles and inhibit the ability to bind to the receptor surface
Caliciviruses	ssRNA	Best activity among 2,080 small molecules screened
HIV-1	ssRNA	TF3 inhibits the entry of the virus by targeting gp41

4.3 Influenza Viruses

The flu is a disease caused by a family of viruses called influenza viruses which belong to the genus *Alphainfluenzavirus* of the virus family *Orthomyxoviridae*. There are four types of influenza viruses: A, B, C, and D. Human influenza A and B viruses cause seasonal epidemics of disease (known as flu season). Current subtypes of influenza A viruses that routinely circulate in people include A (H1N1) and A (H3N2). The influenza type B viruses currently circulating belong to either the B/Yamagata or the B/Victoria lineages. The virus particles can be transmitted from person to person by respiratory droplets, aerosols, and contact. Both influenza A and B viruses can be further classified into clades and sub-clades. Influenza C virus infections generally cause mild illness and are not thought to cause human epidemics. Influenza D viruses primarily affect cattle and are not known to infect or cause illness in people (CDC, 2022; Rawangkan *et al.*, 2021).

4.4 Activity of Green Tea Catechins against the Influenza Viruses

Accumulating evidence confirms the therapeutic effect of polyphenols in various models of influenza virus infection, suggesting that polyphenol-rich plants may be considered a new natural source for the development of future anti-influenza drugs. One of these polyphenol-rich plants is green tea (Rawangkan *et al.*, 2021).

Green tea catechins are potential anti-influenza virus agents. Several research studies have demonstrated that EGCG, the most abundant catechin in green tea, can minimize the infectivity of the influenza A and B viruses. An animal experiment using chickens showed that a diet and water containing green tea components suppressed the replication of the influenza virus (Lee *et al.*, 2012).

Catechins prevent influenza virus replication by blocking hemagglutinin in the adsorption phase. The hemagglutination inhibition assay showed that the galloyl group at the 3-position of EGCG and ECG binds to the hemagglutinin spike on the influenza virus envelope, leading to the inhibition of the attachment activity of the viral hemagglutinin and sialic acid receptor on red blood cells, and then acting on the acidification of the intracellular compartments of endosomes and lysosomes in the penetration and uncoating phases (Rawangkan *et al.*, 2021). EGCG and ECG inhibit the activity of viral RNA and block the viral progeny release phase by inhibiting influenza neuraminidase activity and virus propagation of all influenza virus subtypes including A/H1N1, A/H3N2, and B virus is suppressed, which is similar to the mechanism of action of oseltamivir, an active anti-influenza drug (Song *et al.*, 2005; Rawangkan *et al.*, 2021).

In summary, studies suggest the preventive effects of tea catechins on influenza and common cold. Several experimental in vivo or in vitro studies on catechins and their derivatives have been reported for their anti-influenza virus effects. Three mechanisms have been proposed for their antiviral effect: (1) inhibition of attachment to the host cell, (2) replication inhibition, and (3) neuraminidase (NA) inhibition in the virus (Furushima *et al.*, 2018).

In 2009, the world experienced a swine influenza virus pandemic. In addition, there are concerns about human infection with a highly pathogenic avian influenza virus that is prevalent locally. Furthermore, the prevalence of viruses that acquired resistance to existing NA inhibitors such as oseltamivir (Tamiflu) and zanamivir (Relenza) has also been confirmed. Under such circumstances, the development of new drugs that inhibit virus infection with different mechanisms of action is urgently needed. There are potentials for catechins to serve as effective antiviral therapy with the currently approved drugs (Furushima *et al.*, 2018).

4.5 Coronaviruses; SARS-CoV, MERS-CoV, and COVID-19

Coronaviruses are a large family of viruses that usually cause mild to moderate upper-respiratory tract illnesses, such as the common cold (WHO, 2021). Only seven coronaviruses are known to cause disease in humans. Four of them cause symptoms of the "common cold", and three human coronaviruses cause much more serious lung infections, also called pneumonia (WHO, 2021).

Those three coronaviruses are named SARS-CoV which causes SARS (severe acute respiratory syndrome), MERS-CoV which causes Middle East Respiratory Syndrome (MERS), and SARS-CoV-2 which was identified as the cause of an outbreak of viral pneumonia in Wuhan, China. The disease, later named coronavirus disease 2019 (COVID-19) responsible for COVID-19 (WHO, 2021).

Coronaviruses and influenza viruses cause respiratory diseases. However, they function differently. For example, the genome of the influenza A viruses comprises eight single-stranded RNA molecules, while the typical generic coronavirus genome is a single strand of RNA, 32 kilobases long, and is the largest known RNA virus genome. Figure 31 illustrates the genome of the Influenza virus and coronaviruses responsible for SARS, MERS, and COVID-19 (WHO, 2021). Because those three coronaviruses present the most similarities, COVID-19 research efforts have been building on earlier research done on SARS-CoV and MERS-CoV (Singh *et al.*, 2021).

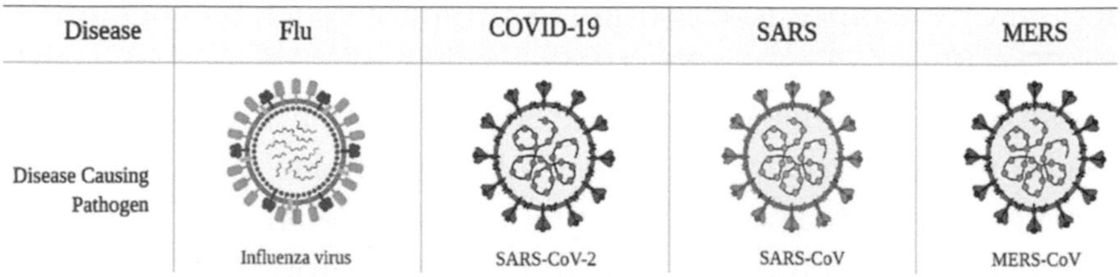

Disease	Flu	COVID-19	SARS	MERS
Disease Causing Pathogen	Influenza virus	SARS-CoV-2	SARS-CoV	MERS-CoV

Figure 31   Structural comparison of the viruses causing respiratory infections (Singh *et al.*, 2021).

4.5.1 Structural Biology of SARS-CoV-2

SARS-CoV-2, the enveloped and spherical-shaped virion with an approximate diameter of 120 nm, contains a positive-sense ssRNA genome of about 30,000 bp. The genome encodes four important structural proteins (nucleocapsid, membrane, envelop, and spike), around 16 non-structural proteins (NSP1–16) and some accessory proteins (Vardhan and Sahoo, 2022).

The trimeric spike (S) glycoprotein protruding from the SARS-CoV-2 envelope plays a key role in viral infection, where the S1 subunit initiates the virus-receptor binding by interacting with the human host cell receptor angiotensin-converting enzyme 2 (ACE2), and the S2 subunit play the role in viral fusion into the target cell. The S1 subunit possesses two key domains, i.e., the N-terminal domain (NTD) and the C-terminal domain (CTD) (Vardhan and Sahoo, 2022).

The CTD of SARS-CoV-2 contains the receptor-binding domain (RBD) that interacts with human ACE2 (Figure 32). Following the interaction of the SARS-CoV-2 virus with the human ACE2 receptor, the host proteases like furin and transmembrane serine protease 2 (TMPRSS2) play the role to activate the S protein that allows the fusion of viral genome into the target cell (Vardhan and Sahoo, 2022).

The human furin enzyme cleaves a specific section to convert the synthesized proteins to biologically active forms. It is also called paired basic amino acid cleaving enzyme (PACE) and subtilisin-like peptidase. It is a calcium-dependent serine endoprotease that cleaves the processing sites of the precursor protein. Furin cleaves the S1 (SPRRARY S) site of SARS-CoV-2, direct emergence in spike glycoprotein activation for viral entry into the host system. It is a potent cleavage site known as the multibasic cleavage site of SARS-CoV-2. In addition, the spike glycoprotein S1/S2 site of SARS-CoV-2 is cleaved by the transmembrane protease serine 2 (TMPRSS2), which played a predominant role in enhancing the spike glycoprotein activity. TMPRSS2 activates the SARS-CoV and SARS-CoV-2 spike glycoprotein through N-terminal and C-terminal cleavage sites, and priming the

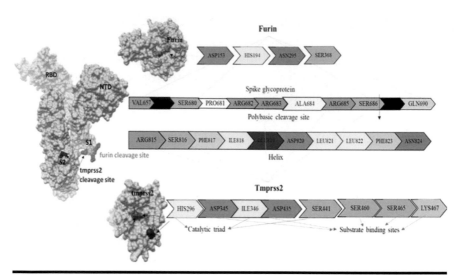

Figure 32 Active sites of furin and TMPRSS2 involved in S1/S2 site cleavage activity of spike protein. The furin catalytic residues are ASP153, HIS194, ASN295, and SER368. For TMPRSS2, the catalytic sites are HIS296, ILE346, and SER441, whereas the substrate binding sites are ASP345, ASP435, SER460, SER465, and LYS467 (Vardhan and Sahoo, 2022).

S protein for virus endocytosis. TMPRSS2 cleaves the S protein in the lungs of SARS-CoV-2 infected person and promotes pathogenicity. Therefore, the TMPRSS2 serine protease opted as an important therapeutic target. In summary, the SARS-CoV-2 S protein S1 site is cleaved by furin, and subsequently, TMPRSS2 mediates the S2 site cleavage and activation of S protein upon interactions with S1/S2 site residues ARG683, ARG685, SER686, ARG815, and SER816 (Figure 32) (Vardhan and Sahoo, 2022).

4.6 Inhibition Effects of Tea Aqueous Extract on SARS-CoV-2

Tea polyphenols possess antiviral activity against various viruses (Ishimoto *et al.*, 2021). Research studies proved that black tea and green tea bag infusion could reduce the viral titer of SARS-CoV-2 (Ishimoto *et al.*, 2021). Aqueous extracts of green tea or black tea inhibit 3C-like protease activity, and thus virus propagation is inhibited and interfered (Ishimoto *et al.*, 2021). The use of tea constituents over synthetic drugs is an exciting treatment option since they are safer, and a higher dose is feasible (Mhatre *et al.*, 2021).

The spike proteins or viral particles of SARS-CoV-2 viruses might be affected by crude tea extracts (i.e., mixtures of polyphenols) rather than polyphenols alone might effectively inactivate SARS-CoV-2. Inactivation of SARS-CoV-2 through synergistic effects, including inhibition of 3C-like protease and interaction with S protein in SARS-CoV-2 by tea polyphenols, should also be considered (Ishimoto *et al.*, 2021).

Catechins retain in human saliva for up to 60 minutes after rinsing with green tea extract (Tsuchiya *et al.*, 1997). The half-life of theaflavins in human saliva is 49–76 minutes after holding freshly brewed black tea in the mouth for 2–5 minutes (Lee *et al.*, 2004). These suggest that even when tea extract is consumed, a sufficient amount of catechins and theaflavins, which are the major components of tea extracts, may remain active in saliva for the necessary time to inactivate SARS-CoV-2. Green tea extract, black tea extract and black tea bag infusion show antiviral activity at lower concentrations than the concentrations usually present in common consuming tea drinks (green tea extract and black tea extract are usually prepared at 2 mg/mL for consumption) (Ishimoto *et al.*, 2021). Drinking black tea and green tea can be expected to decrease the SARS-CoV-2 titer in the mouth and this reduces the viral burden in the mouth and reduces the risk of spreading viruses by droplet transmission from oral sources (Ishimoto *et al.*, 2021).

Susceptibility of the virus to tea leaf extract containing concentrated black tea theaflavins and virucidal catechins has been studied and it seems to be slightly different in the different SARS-CoV-2 strains (Takeda *et al.*, 2021). The susceptibility to tea leaf extract was highest in the Beta, Gamma and Alpha strains but lower in Delta and Kappa strains. Since there are multiple discrepancies in the amino acid sequences of the S protein among these different strains, such amino acid substitutions, especially in the S2 subunit, may impact the sensitivity to tea leaf extract. The S2 subunit contains the fusion peptide, heptapeptide repeat 1 and 2 (HR1, HR2), the transmembrane domain, and the cytoplasm domain. In the strains tested, there were several discrepancies in the amino acid sequence in the HR1 and HR2 regions, but not in the fusion peptide or the transmembrane/cytoplasm domains. As HR1 and HR2 are indispensable to virus envelope-cell membrane fusion and virus entry, these regions may be a possible target of the virucidal compounds contained in tea leaf extract (Takeda *et al.*, 2021). Tea leaf extract inhibits the association of N protein with the RNA genome blocking viral assembly (Takeda *et al.*, 2021; Mhatre *et al.*, 2021; Diniz *et al.*, 2021).

In summary, tea polyphenols have shown promising effects in interrupting transmission, reducing susceptibility, and ameliorating the severity of SARS-CoV, MERS-CoV, and other viral infections. The effects of these micronutrients and bioactive substances on SARS-CoV propose the same effects on SARS-CoV-2 due to their similarities in the phylogenetic and replication cycle (Keflie and Biesalski, 2021). Thus, drinking tea might contribute to the in vivo reduction in the numbers of SARS-CoV-2 virus in the saliva of infected persons, as well as lowering the viral burden in the mouth and upper gastrointestinal and respiratory tracts (Ishimoto *et al.*, 2021).

4.6.1 Inhibition Effects of Catechins on SARS-CoV-2

Catechins are the most abundant bioactive compounds found in green tea (Mhatre *et al.*, 2021). As described earlier, green tea catechins including epigallocatechin gallate (EGCG), epigallocatechin (EGC), epicatechin gallate (ECG), epicatechin (EC), gallocatechin-3-gallate (GCG), gallocatechin (GC), catechin gallate (CG), and catechin (C) are promising antiviral compounds. The antiviral activity of green tea catechins has been shown in numerous single-stranded RNA viruses suggesting the application of tea catechins as a potential treatment for COVID-19 (Furushima *et al.*, 2018; Mhatre *et al.*, 2021; Diniz *et al.*, 2021).

The main protease (Mpro) of SARS-CoV-2, a key component of this viral replication, was considered a prime target for anti-COVID-19 drug development. To find potent Mpro inhibitors, catechins from green tea, were evaluated for inhibition of SARS-CoV-2. Binding affinities and binding modes between these polyphenols and Mpro using *in silico* molecular docking studies were calculated. All eight polyphenols exhibited good binding affinity toward Mpro (−7.1 to −9.0 kcal/mol). However, only epigallocatechin gallate (EGCG), epicatechin gallate (ECG) and gallocatechin-3-gallate (GCG) bind strongly to SARS-CoV-2-Mpro's amino acid residues, His41 and Cys145, that are important for the enzymes' catalytic activity (Ghosh *et al.*, 2020; Zhao *et al.*, 2021; Keflie and Biesalski, 2021; Gogoi *et al.*, 2021; Montone *et al.*, 2021; Ishimoto *et al.*, 2021). These complexes are highly stable, experience less conformational fluctuations and share a similar degree of compactness. The total number of intermolecular H-bond and MM-GBSA analysis affirmed the stability of these three Mpro–polyphenol complexes. Pharmacokinetic analysis additionally suggested that these catechins possess favourable drug-likeness characteristics. These three polyphenols can be used as potential inhibitors against SARS-CoV-2 Mpro and are promising drug candidates for COVID-19 treatment (Figure 33) (Ghosh *et al.*, 2020).

Studies have shown that EGCG inhibited the SARS-CoV main protease (Mpro) or 3 chymotrypsin-like proteases (3CLpro) with an IC50 of 73 µM. Increasing the concentration of (−)-catechin gallate (CG) potentially inhibited SARS-CoV N protein and its association with RNA, with an IC50 of 0.05 µg/m (Figure 34) (Diniz *et al.*, 2021; Mhatre *et al.*, 2021).

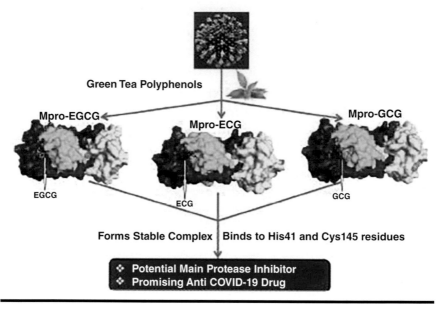

Figure 33   Green tea polyphenols, promising anti Covid-10 drugs.
Abbreviations: COVID-19: coronavirus disease 2019; SARS-CoV-2:
severe acute respiratory syndrome corona virus-2; Mpro: main pro-
tease; MD: molecular dynamics; RMSD: root mean square deviation;
RMSF: root mean square fluctuation; Rg: radius of gyration; SASA:
solvent accessible surface area (Ghosh *et al.*, 2020).

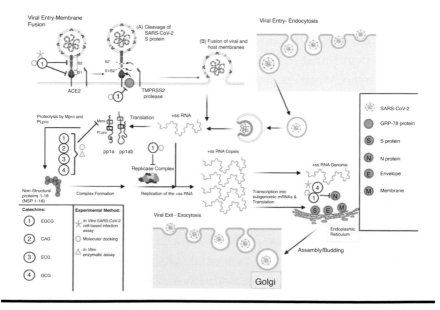

Figure 34   Catechins exert anti-SARS-CoV-2 activities by targeting
different steps of the SARS-CoV-2 lifecycle (Diniz et al., 2021).

Catechins, such as EGCG, bind to the SARS-CoV-2 S protein and inhibit its binding to the ACE2 receptor. EGCG also binds to GRP-78, which potentially blocks its binding to S protein, which may inhibit viral entry. EGCG, ECG, CAG (catechin gallate), and GCG inhibit Mpro of SARS-CoV-2, which blocks viral replication. Molecular docking studies have shown that EGCG binds to RNA-dependent RNA polymerase and other proteins of the replicase complex (NSP6 and NSP15) which may block viral replication. Furthermore, EGCG and GCG bind to and inhibit the association of N protein with the RNA genome blocking viral assembly (Diniz et al., 2021).

4.6.1.1 Inhibition Effects of Epigallocatechin-3-Gallate (EGCG) on SARS-CoV-2

Epigallocatechin gallate (EGCG) has been studied extensively against several viruses as compared to the other catechins owing to its antiviral properties (Mhatre *et al.*, 2021) and is a potential treatment option over synthetic chemical drugs (Mhatre *et al.*, 2021) as a multi-functional bioactive molecule exhibiting several bioactive properties in addition to its antiviral effects (Mhatre *et al.*, 2021). The antiviral properties of this compound against COVID-19 together with other beneficial health effects of this compound could be an advancement in the treatment of this latest pandemic (Mhatre *et al.*, 2021).

Cell membrane receptor angiotensin-converting enzyme 2 (ACE2) is the binding site for the SARS-CoV-2 spike (S) protein, through RBD on the viral membrane and forming RBD–ACE2 complex, by which the SARS-CoV-2 is enabled to get into the host cell where it starts replication. Thus, if a compound exhibits a strong ability to bind the S protein, or has a strong affinity to the ACE2 receptor, which leads to inhibition of RBD–ACE2 complex formation, it would own the potential for repressing viral invading host cells (Wang *et al.*, 2021).

Molecular docking studies of EGCG with possible active binding sites of SARS-CoV-2 revealed that the docking scores and the interactions of the virus with the receptors are a good basis to further explore the application of this molecule in the treatment of COVID-19 (Mhatre *et al.*, 2021). EGCG binds with the viral protein targets with higher affinity than the antiviral drugs, chloroquine and remdesivir (Diniz *et al.*, 2021; Mhatre *et al.*, 2021). The inhibition effects of EGCG on SARS-CoV-2 occur through its actions on the ACE2 receptor, the main protease (Mpro) and RdRp (Figure 35) (Wang *et al.*, 2021). EGCG has a higher atomic contact energy value, binding energy, Ki value, ligand efficiency and surface area than hydroxychloroquine (HCQ) during binding with the spike protein. There are three

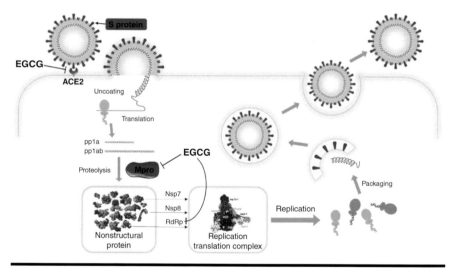

Figure 35   Schematic diagram of SARS-CoV-2 life cycle and the inhibition effects of EGCG. The inhibition effects of EGCG on SARS-CoV-2 replication occur through its actions on the ACE2 receptor, the main protease (Mpro, a 3C-like protease) and RNA-dependent RNA polymerase (RdRp). Abbreviations: EGCG, epigallocatechin-3-gallate; pp1a, nuclear protein phosphatase 1 α; pp1ab, 2′-O-methyltransferase; Mpro, main protease; ACE2, angiotensin-converting enzyme 2; S protein, spike protein; RdRp, RNA-dependent RNA polymerase (Wang *et al.*, 2021).

binding sites on the spike protein. EGCG can bind with all of the three sites, while HCQ binds only with site III, based on the fact that sites I and II are in closer contact with open state location and viral–host contact area. These suggest that EGCG has a stronger ability to inhibit the infection of SARS-CoV-2 to the host cells than HCQ (Wang *et al.*, 2021).

EGCG is an excellent candidate for the next generation of non-toxic compounds that are active against various viral infections with high efficacy, but without the potential to induce viral resistance due to the non-specific binding nature. Therefore, EGCG is suitable to be developed into a novel and potent antiviral compound against many pathogenic viruses to save lives (Hsu, 2015).

A set of experiments evaluated the in vivo distribution of EGCG in human bodies and data showed that the values of EGCG concentration in the colon and intestine were higher than most of the concentrations necessary to promote 3CL protease required to effectively 3CL protease inhibition. More pre-clinical studies, clinical trials and epidemiological analysis will be extremely needed to validate EGCG anti-COVID-19 applications. EGCG and its stable lipophilic derivatives could also be potential prophylactic as well as therapeutic agents looking at their properties to dock at various active sites of SARS-CoV-2. Results from these studies will shed light on the role of the EGCG and the underlying molecular mechanisms for the treatment of SARS-CoV-2 infection (Bimonte *et al.*, 2021).

EGCG could be used as a nutraceutical or dietary supplement, especially in the earlier stages of clinical manifestations of COVID-19. After extensive studies on EGGC regarding its specificity, activity, bioavailability and safety, there can be considerations on the use of this compound in the treatment of viral infections including COVID-19 (Bimonte *et al.*, 2021).

4.6.2 Inhibition Effects of Theaflavins (TFs) on SARS-CoV-2

Theaflavins of black tea include theaflavin (TF1), theaflavin-3-gallate (TF2a), theaflavin-3'-gallate (TF2b) and theaflavin-3,3'-digallate (TF3). The quality of black tea for beneficial health effects partially depends on TF (Takeda *et al.*, 2021). The antiviral activity of TFs has already been reported in numerous single-stranded RNA viruses suggesting the application of these compounds as potential antiviral compounds in treating COVID-19. Among theaflavins, TF3 has been shown to possess better antiviral activity than the other theaflavins (Mhatre *et al.*, 2021). There are some reports of these molecules showing inhibitory effects on SARS-CoV-2. Molecular docking studies of these tea polyphenols on the main protease of SARS-CoV-2 revealed their potential as inhibitors of 3CLpro (Mhatre *et al.*, 2021). In another study, it was shown that TFs blocked the six-helix bundle formation in the glycoprotein 41(gp41) subunit, resulting in the inhibition of virus envelope-cell membrane fusion (Takeda *et al.*, 2021). Studies evaluating the antiviral activity of TFs suggest that the number of galloyl and hydroxyl groups contributes to the virucidal activity of TFs and TF3 is the most potent (Takeda *et al.*, 2021).

4.6.2.1 Inhibition of 3-Chymotrypsin-Like Protease (3CLpro)

Investigations for phytochemical mediated inhibition of SARS-CoV-2 3CLpro, molecular docking and in vitro studies for screening different natural flavonoids have been performed. In vitro investigations, TF1, TF2a, TF2b, and TF3 showed higher potency in 3CLpro inhibition than other catechins including

EGCG at IC50 less than 10 µM. This could be because of the unstable nature of EGCG and the other catechins. Among the TFs, TF2b and TF3 showed better 3CLpro inhibition activity because of the presence of the gallate group (Mhatre *et al.*, 2021). TF2b was found to form more hydrogen bonds with the 3CLpro receptor with a docking score of −9.8 compared to drugs like Lopinavir, Darunavir, and Atazanavir, thus exhibiting the activity to maintain the interaction (Bhardwaj *et al.*, 2020). Another similar study also showed that TF3 had a better docking score (−10.574) than more than 20 antiviral drugs like lopinavir (−9.918), darunavir (−8.843), amprenavir (−8.655) as well as phytochemicals like hesperidin, biorobin, rosmarinic acid, etc. (Peele *et al.*, 2020). Another study described the use of TF in the prophylaxis of SARS-CoV-2. TF2 showed a docking score of −9.8 and TF3 showed a docking score of −10 on the 3CLpro receptor in a molecular docking study (Bhatia *et al.*, 2020).

4.6.2.2 Inhibition of RNA-Dependent RNA Polymerase (RDRP)

Another potential drug target for SARS-CoV-2 is RNA-dependent RNA polymerase (RdRp), which is a key component of the replication machinery of the virus to make multiple copies of the RNA genome (Elfiky, 2021). RdRp in various coronaviruses is remarkably similar. For example, the RdRp of SARS-CoV exhibits ~97% sequence similarity with that of SARS-CoV-2. More importantly, no human polymerase counterpart resembles the sequence/structural homology with RdRp from coronaviruses, and hence, the development of RdRp inhibitors could be a potential therapeutic strategy without the risk of crosstalk with human polymerases (Borgio *et al.*, 2020; Subissi *et al.*, 2014; Zhai *et al.*, 2005).

Because RdRp inhibitors play a crucial role to combat the SARS-CoV-2 infection, a comprehensive molecular docking study with a library of hundred natural polyphenols with potential antiviral properties that may inhibit the SARS-CoV-2 RdRp and prevent the RNA replication was performed. In that study, eight natural polyphenols having binding energy −7.0 kcal/mol or less for molecular dynamics simulation were shortlisted. Further, 150 ns molecular dynamics simulation of RdRp/EGCG, RdRp/TF1, RdRp/TF2a, RdRp/TF2b, RdRp/TF3, RdRp/hesperidin, RdRp/myricetin, and RdRp/quercetagetin, along with RdRp/remdesivir complex was performed and computed the binding energies by the molecular mechanics-Poisson–Boltzmann surface area (MM-PBSA) scheme from last 50 ns trajectories. The study suggests that the complex formation of the SARS-CoV-2 RdRp and eight natural polyphenols is favoured by the intermolecular van der Waals and electrostatic interactions as well as nonpolar solvation free energy. Also, the hotspot residues controlling the receptor-ligand binding were investigated. Finally, molecular dynamics simulation and MM-PBSA study reveals that EGCG, TF2a, TF2b, and TF3 possess a better binding affinity than the control drug remdesivir against the SARS-CoV-2 RdRp. Further, the ADME prediction, toxicity prediction, and target analysis to assess the druggability of the five compounds were searched (Singh *et al.*, 2021).

The results strongly suggest that EGCG, TF2a, TF2b, and TF3 have a stable binding affinity toward RdRp of the SARS-CoV-2 with favourable pharmacokinetic properties. These bioactive compounds exhibit a broad range of therapeutic properties. These four natural polyphenols can act as potential inhibitors for the SARS-CoV-2 RdRp (Tables 4 and 5) (Figure 36) (Singh *et al.*, 2021).

Table 4   Binding Energy (kcal/mol) of the Natural Polyphenols along with the Control Compounds (GTP and Remdesivir) against RdRp of the SARS-CoV-2 (PDB ID: 6M71) by Molecular Docking Study

S. No.	Compound Name	Binding Energy (kcal/mol)	S. No.	Compound Name	Binding Energy (kcal/mol)
1	TF 3	−9.9	52	Cyanidin	−6.3
2	TF 2b	−9.6	53	Daidzein	−6.3
3	TF 1	−9.6	54	Glycitein	−6.3
4	TF 2a	−9.3	55	Wogonin	−6.3
5	Hesperidin	−8.8	56	Phloretin	−6.3
6	EGCG	−7.3	57	Catechin	−6.2
7	Myricetin	−7.2	58	Urolithin B	−6.2
8	Quercetagetin	−7.0	59	Angolensin	−6.2
9	Quercetin	−6.9	60	Pinosylvin	−6.2
10	Curcumin	−6.9	61	Formononetin	−6.2
11	Dihydrorobinetin	−6.8	62	Liquiritigenin	−6.2
12	Peonidin	−6.8	63	Prunetin	−6.2
13	Fisetin	−6.8	64	Alpinetin	−6.2
14	Robinetin	−6.7	65	Biochanin A	−6.2
15	5-Deoxygalangin	−6.7	66	Rhapontigenin	−6.1
16	Kaempferol	−6.7	67	Genistein	−6.1
17	Scutellarein	−6.7	68	Chrysin	−6.1
18	(−)-Epicatechin	−6.7	69	6-Hydroxyflavone	−6.1
19	Purpurin	−6.7	70	Equol	−6.1

(*Continued*)

Table 4 (*Continued*)   Binding Energy (kcal/mol) of the Natural Polyphenols along with the Control Compounds (GTP and Remdesivir) against RdRp of the SARS-CoV-2 (PDB ID: 6M71) by Molecular Docking Study

S. No.	Compound Name	Binding Energy (kcal/mol)	S. No.	Compound Name	Binding Energy (kcal/ mol)
20	Isorhamnetin	−6.7	71	Piceatannol	−6.1
21	Tricetin	−6.6	72	Isorhapontigenin	−6.0
22	Gossypetin	−6.6	73	Resveratrol	−5.8
23	Norathyriol	−6.6	74	Danshensu	−5.7
24	Coumestrol	−6.6	75	Eugenin	−5.6
25	Isosakuranetin	−6.6	76	Sinapic acid	−5.5
26	Pectolinarigenin	−6.6	77	Pterostilbene	−5.5
27	Tangeritin	−6.6	78	Ferulic acid	−5.4
28	Nobiletin	−6.6	79	caffeic acid	−5.4
29	Pratensein	−6.6	80	Isoferulic acid	−5.4
30	Hispidulin	−6.6	81	Dihydrocaffeic acid	−5.4
31	Baicalein	−6.5	82	Gentisic acid	−5.3
32	Apigenin	−6.5	83	Pyrogallol	−5.3
33	Morin	−6.5	84	4-Hydroxycinnamic acid	−5.2
34	Urolithin A	−6.5	85	Resacetophenone	−5.2
35	Acacetin	−6.5	86	Salicydic acid	−5.1
36	Pelargonidin	−6.5	87	Syringic acid	−5.1

(*Continued*)

Table 4 (*Continued*)   Binding Energy (kcal/mol) of the Natural Polyphenols along with the Control Compounds (GTP and Remdesivir) against RdRp of the SARS-CoV-2 (PDB ID: 6M71) by Molecular Docking Study

S. No.	Compound Name	Binding Energy (kcal/mol)	S. No.	Compound Name	Binding Energy (kcal/mol)
37	Irilone	−6.5	88	2-Hydroxybenzoic acid	−5.1
38	Naringenin	−6.5	89	Gallic acid	−5.0
39	Pinocembrin	−6.5	90	3-Hydroxybenzoic acid	−5.0
40	Kaempferide	−6.5	91	4-Hydroxybenzoic acid	−5.0
41	Malvidin	−6.5	92	Vanillin	−5.0
42	Luteolin	−6.4	93	*p*-Coumeric acid	−4.9
43	Dalbergin	−6.4	94	Vanillic acid	−4.8
44	Butein	−6.4	95	Paeonol	−4.8
45	Biochanin A (1-)	−6.4	96	Cinnamic acid	−4.7
46	Fustin	−6.4	97	Protocatechuic acid	−4.6
47	5-Hydroxyflavone	−6.4	98	4-Ethylphenol	−4.5
48	Pinostrobin	−6.4	99	Catechol	−4.5
49	Pinobanksin	−6.4	100	Tyrosol	−4.5
50	Datiscetin	−6.3	101	GTP	−7.9
51	Galangin	−6.3	102	Remdesivir	−7.7

Table 5   Predicted Toxicity Profile of EGCG, TF3, TF2b, TF2a, and Remdesivir

S.No.	Compounds Name	Toxicity Prediction Properties	Predicted Values
1	EGCG	AMES toxicity	No
		Maximum tolerated dose (Human)	0.441 (log mg/ kg/day)
		hERG I inhibitor	NO
		hERG II inhibitor	Yes
		Oral rat acute toxicity (LD)	2.522 (mol/kg)
		Oral rat acute toxicity (LOAEL)	3.065(log mg/ kg_bw/day)
		Hepatotoxicity	No
		Skin sensitivity	No
		T. *pyriformis* toxicity	0.285 (µg/L)
		Minnow toxicity	7.713 log mM
2	TF3	AMES toxicity	NO
		Maximum tolerated dose (Human)	0.438 (log mg/ kg/day)
		hERG I inhibitor	No
		hERG II inhibitor	Yes
		Oral rat acute toxicity (LD)	2.482 (mol/kg)
		Oral rat acute toxicity (LOAEL)	7.443 (log mg/ kg_bw/day)
		Hepatotoxicity	No
		Skin sensitivity	No
		T. *pyriformis* toxicity	0.285 (µg/L)
		Minnow toxicity	9.738 log mM

(Continued)

Table 5 (*Continued*) Predicted Toxicity Profile of EGCG, TF3, TF2b, TF2a, and Remdesivir

S.No.	Compounds Name	Toxicity Prediction Properties	Predicted Values
3	TF 2b	AMES toxicity	No
		Maximum tolerated dose (Human)	0.438 (log mg/kg/day)
		hERG I inhibitor	No
		hERG II inhibitor	Yes
		Oral rat acute toxicity (LD)	2.482 (mol/kg)
		Oral rat acute toxicity (LOAEL)	5.322 (log mg/kg_bw/day)
		Hepatotoxicity	No
		Skin sensitivity	No
		T. *pyriformis* toxicity	0.285 (µg/L)
		Minnow toxicity	8.685 log mM
4	TF 2a	AMES toxicity	No
		Maximum tolerated dose (Human)	0.439 (log mg/kg/day)
		hERG I inhibitor	No
		hERG II inhibitor	Yes
		Oral rat acute toxicity (LD)	2.484 (mol/kg)
		Oral rat acute toxicity (LOAEL)	5.035 (log mg/kg_bw/day)
		Hepatotoxicity	No
		Skin sensitivity	No
		T. *pyriformis* toxicity	0.285 (µg/L)
		Minnow toxicity	4.898 log mM

(*Continued*)

Table 5 (*Continued*)   Predicted Toxicity Profile of EGCG, TF3, TF2b, TF2a, and Remdesivir

S.No.	Compounds Name	Toxicity Prediction Properties	Predicted Values
5	Remdesivir	AMES toxicity	No
		Maximum tolerated dose (Human)	0.15 (log mg/kg/day)
		hERG I inhibitor	No
		hERG II inhibitor	Yes
		Oral rat acute toxicity (LD)	2.043 (mol/kg)
		Oral rat acute toxicity (LOAEL)	1.639 (log mg/kg_bw/day)
		Hepatotoxicity	Yes
		Skin sensitivity	No
		T. *pyriformis* toxicity	0.285 (µg/L)
		Minnow toxicity	0.291 log mM

4.6.2.3 Inhibition of Spike-Receptor-Binding Domain (Spike RBD)

The spike glycoproteins of COVID-19 (similar to SARS-CoV) attach to the angiotensin-converting enzyme (ACE2) and transit over a stabilized open state for the viral internalization to the host cells and propagate with great efficacy. Higher rate of mutability makes this virus unpredictable/less sensitive to protein-/nucleic acid-based drugs. In this emergent situation, drug-induced destabilization of spike binding to RBD could be a good strategy (Maiti and Banerjee, 2021).

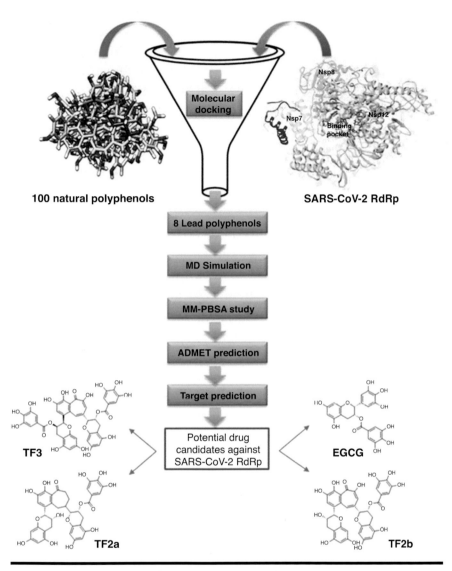

Figure 36 Flow chart of the methodology for shortlisting the best natural polyphenolic inhibitor of the SARS-CoV-2 RdRp (Singh *et al.*, 2021).

Studies suggest that TFs have a good binding with the RBD in SARS-CoV-2 with a high docking score. The driving forces of these interactions were majorly hydrophobic interactions along with hydrogen bonds at sites like ARG454, PHE456, ASN460, CYS480, GLN493, ASN501, and VAL503 of SARS-CoV-2 RBD, near the ACE2-S protein contact area. For the most favourable interaction of TF with the RBD in SARS-CoV-2, the binding energy (ΔG) was estimated to be −8.53 kcal/mol (Lung *et al.*, 2020).

In the investigations that were performed by bioinformatics (CASTp: computed atlas of surface topography of protein, PyMol: molecular visualization) and molecular docking (PatchDock and Autodock), it was revealed that tea catechins, namely EGCG, theaflavin gallate and Theaflavin-3,3'-digallate (TF3), are the most potent molecules, which could be used for the COVID-19 treatment as such or after some chemical modification by making it more efficient in interfering the transition between open and closed states of the viral spike. These molecules demonstrated higher atomic contact energy (ACE) value, binding energy, Ki value, ligand efficiency, surface area and more amino acid interactions than hydroxychloroquine (HCQ) during binding in the central channel of the spike protein. Moreover, out of three distinct binding sites (I, II, and III) of spike core when HCQ binds only with site III (farthest from the nCoV-RBD of ACE2 contact), EGCG and theaflavin gallate bind all three sites. As sites I and II are in closer contact with open state locations and viral–host contact areas,

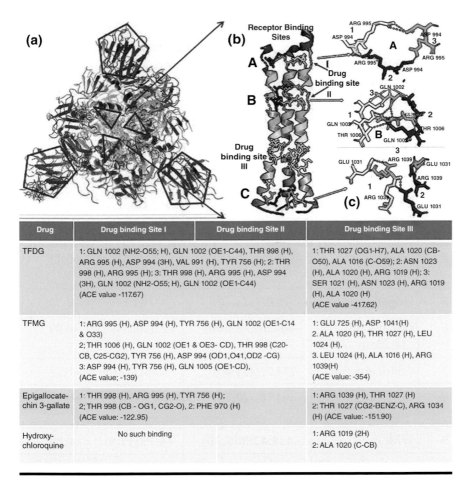

Drug	Drug binding Site I	Drug binding Site II	Drug binding Site III
TFDG	1: GLN 1002 (NH2-O55; H), GLN 1002 (OE1-C44), THR 998 (H), ARG 995 (H), ASP 994 (3H), VAL 991 (H), TYR 756 (H); 2: THR 998 (H), ARG 995 (H); 3: THR 998 (H), ARG 995 (H), ASP 994 (3H), GLN 1002 (NH2-O55; H), GLN 1002 (OE1-C44) (ACE value -117.67)		1: THR 1027 (OG1-H7), ALA 1020 (CB-O50), ALA 1016 (C-O59); 2: ASN 1023 (H), ALA 1020 (H), ARG 1019 (H); 3: SER 1021 (H), ASN 1023 (H), ARG 1019 (H), ALA 1020 (H) (ACE value -417.62)
TFMG	1: ARG 995 (H), ASP 994 (H), TYR 756 (H), GLN 1002 (OE1-C14 & O33) 2; THR 1006 (H), GLN 1002 (OE1 & OE3- CD), THR 998 (C20-CB, C25-CG2), TYR 756 (H), ASP 994 (OD1,O41,OD2 -CG) 3: ASP 994 (H), TYR 756 (H), GLN 1005 (OE1-CD), (ACE value; -139)		1: GLU 725 (H), ASP 1041(H) 2. ALA 1020 (H), THR 1027 (H), LEU 1024 (H), 3. LEU 1024 (H), ALA 1016 (H), ARG 1039(H) (ACE value: -354)
Epigallocate-chin 3-gallate	1: THR 998 (H), ARG 995 (H), TYR 756 (H); 2; THR 998 (CB - OG1, CG2-O), 2: PHE 970 (H) (ACE value: -122.95)		1: ARG 1039 (H), THR 1027 (H) 2: THR 1027 (CG2-BENZ-C), ARG 1034 (H) (ACE value: -151.90)
Hydroxy-chloroquine	No such binding		1: ARG 1019 (2H) 2: ALA 1020 (C-CB)

Figure 37 Homotrimer coronavirus spike protein and its probable ligand binding site (a). Hydrogen bond stabilizing the central core of coronavirus spike protein (b). Different ligand-binding efficiencies in different drug-binding sites with ACE value. When ECGC and TG interact three different drug-binding sites involving with more number of amino acids the HCQ interacts only on-site III which is far from RBD involving with less number of amino acids. TFMG stands for theaflavin gallate and TFDG stands for theaflavin -3,3′-digallate (Maiti and Banerjee, 2021).

Table 6   Comparative Analysis of HCQ, Catechin, Catechin Gallate, Epicatechin 3-O-Gallate, Epigallocatechin, Epigallocatechin 3-Gallate, Gallocatechin, Gallocatechin Gallate, TFMG and TFDG Binding with ACE2 and nCoV2 through MD Using Patchdock and Autodock

S. No.	Compound	Patchdock ACE Value with nCoV2 (area in A²)	Binding Energy (Ki value in μmol)	Ligand Efficiency
1	HCQ	−293.32 (616.90)	−3.64(2.14)	−0.16
2	Catechin	−266.41(525.20)	−4.23(799.34)	−0.2
3	Catechin gallate	−393.05(732.90)	−4.16(885.65)	−0.13
4	Epicatechin 3-O-gallate	−308.25(689.60)	−3.55(2.51)	−0.11
5	Epigallocatechin	−270.01(523.40)	−3.94(1.29)	−0.18
6	Epicatechin 3-gallate	**−407.58(723.90)**	−2.98(6.54)	−0.09
7	Gallocatechin	−274.72(471.40)	−4.76(322.09)	−0.22
8	Gallocatechin gallate	−364.16(722.70)	−3.79(1.67)	−0.11
9	TFMG	**−434.42(906.20)**	**−6.72(11.9)**	−0.13
10	TFDG	**−465.17(1034.60)**	−1.85(44.19)	−0.06

Note: Bold values are highly significant rather than low significant value in HCQ (−293.32).

TFMG stands for theaflavin gallate and TFDG stands for theaflavin -3,3′-digallate.

these drugs might have significant effects. Considering the toxicity and side effects of chloroquine/HCQ, these molecules are important drugs (Figure 37, Table 6) (Maiti and Banerjee, 2021).

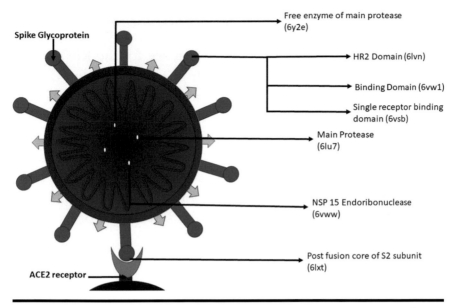

Figure 38   Sites on SARS-CoV-2 docked with therapeutic molecules (Mhatre *et al.*, 2021).

4.6.2.4 *Inhibition of Structural Proteins*

Molecular docking studies for phytochemicals which have been already reported for their antiviral activity are being performed with different proteins of SARS-CoV-2 shown in Figure 38. These docking results are compared with two drugs which are possibly the most potent against COVID-19, remdesivir and chloroquine. The protein data bank IDs of the main protease, HR2 domain, post-fusion core of the S2 subunit, S protein, RBD–ACE2 complex, NSP15 endoribonuclease, and free enzyme crystal structure main protease are 6lu7, 6lvn, 6lxt, 6vsb, 6vw1, 6vww, and 6y2e, respectively (Mhatre *et al.*, 2021). In one such study performed on 169 different phytoconstituents typically used as spices or flavours, TF3 was found to have a considerable affinity with M protease along with antioxidant activity (Zhang *et al.*, 2020).

4.6.2.5 Inhibition of ACE2 Receptor-Binding Ability

Since the ACE2 receptor is the attaching site for the RBD of the viral S protein, direct interaction with ACE2 can prevent the infection of the cell by the virus. In a study, TF3 was found to directly bind to the ACE2 receptor, thus encouraging its use for prophylaxis (Zhang *et al.*, 2020). The ability of TF3 to prevent the spike RBD from attaching to the ACE2 receptor was also shown by Maiti and Banerjee (2021). Figure 39 shows the possible mechanisms of action for tea polyphenols on different active sites. Table 7 summarizes the studies performed on SARS-CoV-2 using tea polyphenols (Mhatre *et al.*, 2021).

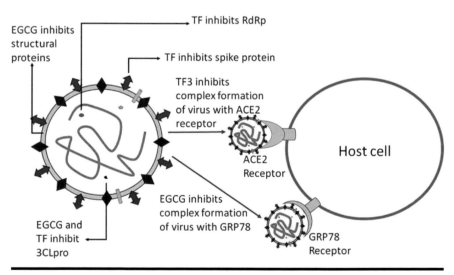

Figure 39   Depiction of the role of tea polyphenols on druggable targets of COVID-19 (Mhatre *et al.*, 2021).

Table 7   Antiviral Activity of Tea Polyphenols on COVID-19 Receptors (Mhatre *et al.*, 2021)

Compound	Receptor Target	Activity
EGCG	3CLpro	Inhibition activity is better than other phytochemicals
TF2b	3CLpro	Better interaction as compared to repurposed drugs
TF	RdRp	Best interaction than compounds from Chinese traditional medicine
TF	Spike RBD	Good molecular docking score with multiple hydrogen bonds
TF3	M protease	Better affinity among 169 phytoconstituents used as spices and flavours
EGCG	Multiple structural proteins	Better interaction than Chloroquine and Remdesivir
TF3	ACE2 Receptor	Directly binds to the receptor and acts as a prophylactic agent
TF2,TF3	3CLpro	Molecular docking scores of -9.8 and -10 on the receptor respectively
TF	3CLpro	Better activity than 24 approved antiviral drugs and screened phytochemicals

4.7 Challenges and Future Perspectives

COVID-19 is spreading at an alarming rate, and the lack of an approved treatment is causing a major load on the health-care systems. Several antiviral drugs are under clinical trials; however, owing to possible side effects, higher doses of these drugs cannot be administered. Compared with the mechanisms of action of possible drug candidates in previously known viral diseases, some potential viral targets and drugs that can act on these sites can be shortlisted. EGCG and TFs are polyphenolic catechins found abundantly in green tea and black tea, respectively, with a vast array of health benefits. Their antiviral activities have also been reported against various viral infections. An in-depth analysis of the antiviral activities of EGCG and TFs reveals that both of them are wide-spectrum antiviral molecules. They act at different stages of the viral cycle. Some studies have also suggested that EGCG and TFs have prophylactic activity. 3CLpro is a vital enzyme found in SARS-CoV and SARS-CoV-2. Considering the genomic similarities between the two viruses, the nature of 3CLpro found in them is also similar. 3CLpro is also a druggable target owing to its important function in viral cell replication. EGCG and all the TFs are found to inhibit 3CLpro at lesser concentrations and also exhibit good binding to this target. TFs also exhibit a good binding with the RBD in SARS-CoV-2, by forming hydrophobic interactions as well as hydrogen bonds with some sites on viral S protein. Additionally, TFs are known to show RdRp inhibition and ACE2 binding activity. Hence, both EGCG and TFs are potential antiviral agents which should be explored as treatment and prophylactic alternatives for COVID-19. GRP78 is a protein found in ER and performs the role of preventing the unfolding of the proteins that are being synthesized. When the cell is under external stress, GRP78 moves to the cell membrane from the ER and becomes a target for viral infection. EGCG is known to interact with the COVID-19 S protein GRP78 binding site, which is a druggable target, via hydrogen and non-bonded interactions, thus exhibiting its antiviral activity (Mhatre *et al.*, 2021). Apart from these, broad-spectrum antiviral activity has already been established for these tea polyphenols. Hence, among all the suggested targets for COVID-19, tea polyphenols can potentially show inhibitory activity. Only further docking and modelling studies can help analyze these interactions in detail. However, EGCG is unstable and when consumed orally has low bioavailability. It tends to get oxidized quite easily before it reaches the target. Many studies have suggested structural derivatives of EGCG to enhance its bioavailability. Ester derivatives of EGCG showed better antioxidant activity in scavenging reactive oxygen species like peroxyl radicals. The antiviral activity was also increased since the derivatives showed better inhibition of HCV protease as well as a-glucosidase in HIV treatment (Zhong *et al.,* 2011). Some studies suggest the use of nanoparticles to encapsulate EGCG to attain better efficacy (Munin and Edwards-Lévy, 2011). In a study involving HCV, the combination of EGCG with monoclonal antibodies showed improved delivery of EGCG in vivo. Lipophilic derivatives of water-soluble EGCG like EGCG palmitate and EGCG stearate were evaluated for improvements in antimicrobial properties, and EGCG palmitate was found to be 8.7 times more potent in HSV infection than EGCG. The same author also suggested the prophylactic use of EGCG palmitate by incorporating it into sanitizers. EGCG-monoesters were also synthesized with butanoyl, octanoyl, laurinol, palmitoyl, and eicosanoyl modifications to test the effects of alkyl length on the anti-influenza activity of EGCG. It was observed that long acyl groups could significantly increase the anti-influenza activity of EGCG, and EGCG palmitate, the most potent among all the ester derivatives of EGCG, was about 24 times more potent than native EGCG. EGCG stearate was shown to inhibit HSV infection as well as treat its associated symptoms. By performing substitution like methylation, acylation, esterification, or glycosylation at different sites of EGCG, the pharmacokinetic and pharmacodynamic properties of EGCG can be improved. EGCG and its derivatives show some promising results in multiple viral infections and it will be worth studying the applications of these molecules in COVID-19 with the help of molecular docking (Mhatre *et al.*, 2021).

Both the tea polyphenols need further detailed evaluation to validate their anti-COVID-19 applications. TFs, especially TF2b and TF3, can be used as good prophylactic agents owing to their ability to bind to RBD. EGCG and its stable lipophilic derivatives could also be potential prophylactic as well as therapeutic agents looking at their properties to dock to various active sites of SARS-CoV-2 (Mhatre *et al.*, 2021).

Tea is the most widely consumed beverage in the world, and developing antiviral polyphenolic molecules from the same is an exciting idea. The FDA has already assigned these polyphenols the coveted GRAS status, which further encourages their use over synthetic toxic antiviral drugs concerning higher dosages. However, concluding EGCG and TFs as drug candidates based on currently available literature would be an overstatement. Since these polyphenols have no specific activity, we cannot be sure of their targeted activity to the COVID-19 receptors discussed earlier. These might also bind to other proteins in the body and produce side effects. Establishing compliance with the stringent regulatory affairs for molecules to be deemed drugs over the same molecule as nutraceutical is time-consuming. Hence, in the immediate future in terms of application in COVID-19, these molecules may not be used in treatment, but as dietary supplements or nutraceuticals. After extensive studies on these polyphenols regarding their specificity, activity, bioavailability, and safety, there can be considerations on their use in the treatment of viral infections including COVID-19 (Mhatre *et al.*, 2021).

For better bioavailability and interaction with different viral components, spike-glycoprotein nanoparticulated catechin derivatives may be used alone or in combination with other bioactive peptides/chitosan/lectins (Hu *et al.*, 2012).

Abbreviations

3CLpro	3 chymotrypsin-like protease
ACE2	receptor–angiotensin converting enzyme II
Al	aluminium
AP-1	activator protein-1
C	catechin
Ca	calcium
Cd	cadmium
CG	catechin gallate
CHIKV	Chikungunya virus
COVID-19	coronavirus disease 2019
Cr	chromium
CRC	colorectal cancer
CTD	C-terminal domain
Cu	copper
CVDs	cardiovascular diseases
DENV	Dengue Virus
DNA	deoxyribonucleic acid
EC	epicatechin
ECG	epicatechin gallate
EGC	epigallocatechin
EGCG	epigallocatechin gallate
ERK	extracellular signal-regulated kinase
F	fluorine
Fe	Iron

GC	gallocatechin
GCG	gallocatechin-3-gallate
GRP78	glucose regulated protein-78
H2O2	hydrogen peroxide
HBV	hepatitis B Virus
HCQ	hydroxychloroquine
HCV	Hepatitis C Virus
HIV	human immunodeficiency virus
HR1	heptapeptide repeat 1
HR2	heptapeptide repeat 2
HSV	Herpes Simplex Virus
JNK	c-Jun NH2-terminal Kinase
K	potassium
MAP kinase	mitogen-activated protein kinase
Mg	magnesium
MM-PBSA	molecular mechanics-Poisson-Boltzmann surface area
Mn	manganese
Mpro	main protease
NA	neuraminidase
Na	sodium
NAFLD	non-alcoholic fatty liver disease
Ni	nickel
NK	natural killer

References

Abudureheman, B.; Yu, X.; Fang, D.; Zhang, H. Enzymatic oxidation of tea catechins and its mechanism. *J. Molecules* 2022, 27, 942.

Afaq, F.; Katiyar, S.K. Polyphenols: Skin photoprotection and inhibition of photocarcinogenesis. *J. Mini Rev. Med. Chem.* 2011, 11(14), 1200–1215.

Ahmad, N.; Feyes, D.K.; Nieminen, A.L.; Agarwal, R.; Mukhtar, H. Green tea constituent epigallocatechin-3-gallate and induction of apoptosis and cell cycle arrest in human carcinoma cells. *J. Natl. Cancer Inst.* 1997, 89(24), 1881–1886.

Ahmad, N.; Katiyar, S.K.; Mukhtar, H. Cancer chemoprevention by tea polyphenols. In: *Nutrition and Chemical Toxicity*. C. Ioannides (ed.), West Sussex, England: John Wiley &Sons Ltd., 1996, 301–343.

Aiyegoro, O.A.; Okoh, A.I. Phytochemical screening and polyphenolic antioxidant activity of aqueous crude leaf extract of helichrysum pedunculatum. *Int. J. Mol. Sci.* 2009, 10(11), 4990–5001.

American Diabetes Association. Diagnosis and classification of diabetes mellitus. *J. Diabetes Care* 2009, 32(Suppl 1), S62–S67.

Andersen, L.F.; Jacobs, D.R., Jr; Carlsen, M.H.; Blomhoff, R. Consumption of coffee is associated with reduced risk of death attributed to inflammatory and cardiovascular diseases in the Iowa women's health study. *Am. J. Clin. Nutr.* 2006, 83, 1039–1046.

Apak, R.; Güçlü, K.; Demirata, B.; Ozyürek, M.; Celik, S.E.; Bektaşoğlu, B.; Berker, K.I.; Ozyurt, D. Comparative evaluation of various total antioxidant capacity assays applied to phenolic compounds with the CUPRAC assay. *J. Mol.* 2007, 12, 1496–1547.

Arts, I.C.W. A review of the epidemiological evidence on tea, flavonoids, and lung cancer. *J. Nutr.* 2008, 138, 1561S–1566S.

Balentine, D.A. Manufacturing and chemistry of tea. In: *Phenolic Compounds in Food and Their Effects on Health I, ACS Symposium Series*, Vol. 506, Chapter 8, M-T Huang, C-T Ho und Chang, Y. Lee (eds.), Washington, DC: American Chemical Society, 1992, 102–117.

Banerjee, B.; Chaudhuri, T.C. *Therapeutic Effects of Tea.* Cambridge: CRC, 2005, 220.

Barbosa, D.S. Green tea polyphenolic compounds and human health. *J. Verbr. Lebensm.* 2007, 2, 407–413.

Behl, C. Oxidative stress in Alzheimer's disease: Implications for prevention and therapy. *Alzheimer's Disease* 2005, 38, 65–78.

Bennett, M.R. Reactive oxygen species and death: Oxidative DNA damage in atherosclerosis. *J. Circ. Res.* 2001, 88, 648–650.

Bharadwaz, A., Bhattacharjee, C. Extraction of polyphenols from dried tea leaves. *J. Sci. Eng. Res.* 2012, 3(5), 1–5.

Bhardwaj, V.K.; Singh, R.; Sharma, J.; Rajendran, V.; Purohit, R.; Kumar S. Identification of bioactive molecules from tea plant as SARS-CoV-2 main protease inhibitors. *J. Biomolecular Struct. Dyn.* 2020, 39(10), 1–13.

Bhatia, S.; Giri, S.; Lal, A.F.; Singh, S. *Battle against Coronavirus: Repurposing Old Friends (Food Borne Polyphenols) for New Enemy (COVID-19).* ChemRxiv. Cambridge: Cambridge Open Engage, 2020.

Bimonte, S.; Forte, C.A.; Cuomo, M.; Esposito, G.; Cascella, M.; Cuomo, A. An overview on the potential roles of EGCG in the treatment of COVID-19 infection. *J. Drug Des. Devel. Ther.* 2021, 15, 4447–4454

Biswas, S.; Bhattacharyya, J.; Dutta, A.G. Oxidant induced injury of erythrocyte—Role of green tea leaf and ascorbic acid. *J. Mol. Cell Biochem.* 2005, 276, 205–210.

Blot, W.J.; Chow, W.H.; McLaughlin, J.K. Tea and cancer: A review of the epidemiological evidence. *Eur. J. Cancer. Prev.* 1996, 5, 425–438.

Boldogh, I.; Kruzel, M.L. Colostrinin: An oxidative stress modulator for prevention and treatment of age-related disorders. *J. Alzheimers Dis.* 2008, 13(3), 303–321.

Borah, P.; Handique, P.J. Identification of potential plantbased inhibitor against viral proteases of SARSCoV2 through molecular docking, MMPBSA binding energy calculations and molecular dynamics simulation. *J. Mol. Divers.* 2021, 25(3), 1963–1977

Borgio, J.F.; Alsuwat, H.S.; Al Otaibi, W.M.; Ibrahim, A.M.; Almandil, N.B.; Al Asoom, L.I.; Salahuddin, M.; Kamaraj, B.; AbdulAzeez, S. State-of-the-art tools unveil potent drug targets amongst clinically approved drugs to inhibit helicase in SARS-CoV-2. *J. Arch. Med. Sci.* 2020, 16(3), 508–518.

Bosch, B, J.; Zee, R.V.D.; Haan, C.A.M.; Rottier, J.M.J.P. The coronavirus spike protein is a class I virus fusion protein: Structural and functional characterization of the fusion core complex. *J. Virol.* 2003, 77(16), 8801–8811.

Brewer, G.J. Risks of copper and iron toxicity during aging in humans. *J. Chem. Res. Toxicol.* 2010, 23, 319–326.

Cao, Y.S.; Zhao, C.N.; Gan, R.Y.; Xu, X.Y.; Wei, X.L.; Croke, H.; Atanasov, A.G.; Li, H.B. Effects and mechanisms of tea and its bioactive compounds for the prevention and treatment of cardiovascular diseases. *J. Antioxidants (Basel)* 2019, 8(6), 166.

Cavaliere, C.; Cucci, F.; Foglia, P.; Guarino, C.; Samperi, R.; Laganà, A. Flavonoid profile in soybeans by high-performance liquid chromatography/tandem mass spectrometry. *J. Rapid Commun. Mass Spectrom.* 2007, 21(14), 2177–2187.

CDC. *Centre for Disease Control and Prevention, Types of Influenza Viruses.* 2022. www.cdc.gov/flu/about/viruses/types.htm

Chang, Q.K.; Chen, Z.M. *Tea and Health.* Beijing, China: Press of Chinese Agricultural Sciences, 1994.

Chanwitheesuk, A.; Teerawutgulrag, A.; Rakariyatham, N. Screening of antioxidant activity and antioxidant compounds of some edible plants of Thailand. *J. Food Chem.* 2005, 92, 491–497.

Chung, J.Y.; Huang, C.; Meng, X.; Dong, Z.;Yang, C.S. Inhibition of activator protein 1 activity and cell growth by purified green tea and black tea polyphenols in H-ras-transformed cells. Structure-activity relationship and mechanisms involved. *J. Cancer Res.* 1999, 59(18), 4610–4617.

Ciesek, S.; Hahn, T.V.; Colpitts, C.C.; Schang, L.M.; Friesland, M.; Steinmann, J.; Manns, P.M.; Ott, M.; Wedemeyer, H.; Meuleman, P.; Steinmann, E. The green tea polyphenol, epigallocatechin-3-gallate, inhibits hepatitis C virus entry. *J. Hepatology* 2011, 54(6), 1947–1955.

Crozier, A.; Clifford, M.N.; Ashihara, H. *Plant Secondary Metabolites: Occurrence, Structure and Role in the Human Diet.* Oxford: Wiley-Blackwell, 2006.

Crozier, A.; Jaganath, I.B.; Clifford, M.N. Dietary phenolics: Chemistry, bioavailability and effects on health. *J. Nat. Prod. Rep.* 2009, 26(8), 1001–1043.

Darley-Usmar, V.; Halliwell, B. Blood radicals: Reactive nitrogen species, reactive oxygen species, transition metal ions, and the vascular system. *J. Pharm. Res.* 1996, 13, 649–662.

Dias, R., Brás, N., Fernandes I., Pérez-Gregorio M., Mateus N., Freitas V. Molecular insights on the interaction and preventive potential of epigallocatechin-3-gallate in celiac disease. *Int. J. Biol. Macromol.* 2018, 112, 1029–1037.

Ding, S.; Jiang.H; Fang.J. Regulation of immune function by polyphenols. *J. Immun. Res.* 2018, 2018(3), 1–8.

Diniz, L.R.L.; Elshabrawy, H.A.; Souza, M.T.S.; Duarte, A.B.S.;Datta, S.; Sousa, D.P. Catechins: Therapeutic perspectives in COVID-19-associated acute kidney injury. *J. Molecules* 2021, 26(19), 5951.

Dou J.; Lee, V.S.; Tzen, J.T.; Lee, M.R. Identification and comparison of phenolic compounds in the preparation of Oolong tea manufactured by semifermentation and drying processes. *J. Agric. Food chem.*, 2007, 55(18), 7462–7468.

Dufresne, C.J.; Farnworth, E.R. A review of latest research finding on the health promotion properties of tea. *J. Nutr. Biochem.* 2001, 12(7), 404–421.

Elfiky, A.A. SARS-CoV-2 RNA dependent RNA polymerase (RdRp) targeting: An in-silico perspective. *J. Biomol. Struct. Dyn.* 2021, 39(9), 3204–3212.

Ellis, R.; Nyirenda, H.E. A successful plant improvement programme on tea (Camellia sinensis). *Exp. Agric.* 1995, 31(3), 307–323.

Fabre, N.; Rustan, I.; Hoffmann, E.; Quetin-Leclercq, J. Determination of flavone, flavonol, and flavanone aglycones by negative ion liquid chromatography electrospray ion trap mass spectrometry. *J. Am. Soc. Mass Spectrom.* 2001, 12(6), 707–715.

Falcão, L.; Araújo. M.E.M. Vegetable tannins used in the manufacture of historic leathers. *J. Mol.* 2018, 23(5), 1081.

Falcão, L.; Araújo. M.E.M. Tannins characterisation in new and historic vegetable tanned leathers fibres by spot tests. *J. Cult. Herit.* 2011, 12 (2) 149–156

Fan, W.; Tezuka, Y.; Komatsu, K.; Namba, T.; Kadota, S.; Prolyl endopeptidase inhibitors from the underground part of Rhodiola sachalinensis. *J. Biol. Pharm Bull.* 1999, 22, 157–161.

Fatima, A.; Alam, A.; Singh, R. Therapeutic potential of phytoestrogens. In: V. Rani, U. Yadav (eds.), *Functional Food and Human Health*. Singapore: Springer, 2018, 297–327.

Feng, R.; Lu, Y.; Bowman, L.L.; Qian, Y.; Castranova, V.; Ding, M. Inhibition of activator protein-1, NF-Kappab, and Mapks and induction of phase 2 detoxifying enzyme activity by chlorogenic acid. *J. Biol. Chem.* 2005, 280(30), 27888–27895.

Ferguson, L.R. Role of plant polyphenols in genomic stability. *J. Mutat. Res.* 2001, 475(1–2), 89–111.

Furushima, D.; Ide, K.; Yamada, H. Effect of tea catechins on influenza infection and the common cold with a focus on epidemiological/clinical studies. *J. Mol.* 2018, 23(7), 1795.

Garg, A.; Garg, S.; Zaneveld, L.J.; Singla, A.K. Chemistry and pharmacology of the citrus bioflavonoid hesperidin. *J. Phytother. Res.* 2001, 15(8), 655–669.

Gerald, W.; Song, M.; Clain, M.C. Polyphenols and gastrointestinal diseases. *J. Curr. Opin. Gastroenterol.* 2006, 22(2), 165–170.

Ghosh, R.; Chakraborty, A.; Biswas, A.; Chowdhuri, S. Evaluation of green tea polyphenols as novel corona virus (SARS CoV-2) main protease (Mpro) inhibitors – an in silico docking and molecular dynamics simulation study. *J. Biomol. Struct. Dyn.* 2020, 39(12), 4362–4374.

Gogoi, B.; Chowdhury, P.; Goswami, N. Identification of potential plant-based inhibitor against viral proteases of SARS-CoV-2 through molecular docking, MM-PBSA binding energy calculations and molecular dynamics simulation. *J. Mol Divers.* 2021, 25, 1963–1977.

Gothandam, K.; Ganesan, V.S.; Ayyasamy, T.; Ramalingam, S. Antioxidant potential of theaflavin ameliorates the activities of key enzymes of glucose metabolism in high fat diet and streptozotocin—Induced diabetic rats. *J. Redox Rep.* 2019, 24(1), 41–50.

Greca, M.; Zarrelli, A. Nutraceuticals and Mediterranean diet. *J. Med. Aromat. Plants* 2012, 1(6), e126.

Grijalba, M.T.; Andrade, P.B.; Meinicke, A.R.; Castilho, R.F.; Vercesi, A.E.; Schreier, S. Inhibition of membrane lipid peroxidation by a radical scavenging mechanism: A novel function for hydroxyl-containing ionophores. *J. Free Radic. Res.* 1998, 28(3), 301–318.

Grotewold, E. *The Stereochemistry of Flavonoids. Science of Flavonoids*. New York: Springer, 2006, 1.

Gunter, G. C. K.; Dell'Aquila,C. ; Aspinall, S.M. ; Runswick, S.A. ; Mulligan, A.A.; Bingham,S.A. Phytoestrogen content of beverages, nuts, seeds, and oils. *J Agric Food Chem.* 2008, 27, 56(16), 7311–7315.

Haider, N.; Larose, L. Harnessing adipogenesis to prevent obesity. *J. Adipocyt.* 2019, 8(1), 98–104.

Hammett-Stabler, C.A. *Osteoporosis from Pathophysiology to Treatment: Special Topics in Diagnostic Testing.* Washington, DC: AACC Press, 2004, ISBN 1-59425-005-7.

Hampton, M.G. Production of Black Tea. In: *Tea: Cultivation to Consumption.* K.C. Willson, M.N. Clifford (eds.), London: Chapman and Hall, 1992, 459–511.

Han, X.; Shen, T.; Lou, H. Dietary polyphenols and their biological significance. *Int. J. Mol. Sci.* 2007, 8, 950–988.

Hashimoto, F.; Ono, M. Evaluation of the anti-oxidative effect (in vitro) of tea polyphenols. *J. Biosci. Biotechnol. Biochem.* 2003, 67(2), 396–401

Haslam, E. Thoughts on thearubigins. *J. Phytochem.* 2003, 64, 61–73.

Haubera, I.; Hohenberga, H.; Holstermanna, B.; Hunsteinb, W.; Hauber, J. The main green tea polyphenol epigallocatechin-3-gallate counteracts semen-mediated enhancement of HIV infection. *Proc. Natl. Acad. Sci. U S A,* 2009, 106(22), 9033–9038.

Hegarty, V.M.; May, H.M.; Khaw, K.T. Tea drinking and bone mineral density in older women. *Am. J. Clin. Nutr.* 2000, 71, 1003–1007.

Hensley, K.; Robinson, K.A.; Gabbita, S.P.; Salsman, S.; Floyd, R.A. Reactive oxygen species, cell signaling, and cell injury. *J. Free Radic. Biol. Med.* 2000, 28(10), 1456–1462.

Hodek, P.; Trefil, P.; Stiborová, M. Flavonoids-potent and versatile biologically active compounds interacting with cytochromes P450. *J. Chem. Biol. Interact.* 2002, 139(1), 1–21.

Hollman, P.C.; Katan, M.B. Dietary flavonoids: Intake, health effects and bioavailability. *J. Food Chem. Toxicol.* 1999, 37, 937–942.

Hoshiyama, Y.; Kawaguchi, T.; Miura, Y. A prospective study of stomach cancer death in relation to green tea consumption in Japan. *Br. J. Cancer* 2002, 87(3), 309–313.

Hsu, S. Compounds derived from epigallocatechin-3-gallate (EGCG) as a novel approach to the prevention of viral infections. *J. Inflamm. Allergy Drug Targets* 2015, 14(1), 13–18.

Hu, B.; Ting, Y.; Yang, X.; Tang, W.; Zeng, X., Huang, Q. Nanochemoprevention by encapsulation of (−)-epigallocat-echin-3-gallate with bioactive peptides/chitosan nanoparticles for enhancement of its bioavailability. *J. Chem. Commun.* 2012, 48(18), 2421–2423.

Hu, J.G.; Zhang, L.J.; Sheng, Y.Y.; Wang, K.R. Screening tea hybrid with abundant anthocyanins and investigating the effect of tea process-ing on foliar anthocyanins in tea. *J. Folia Hort.* 2020, 32(2), 1–12.

Huang, H.T. *Science and Civilization in China.* Vol. 6, Biology and Biological Technology, Part 5, Fermentations and Food Science, Cambridge University Press, 2001, ISBN-13 978-0521652704.

Huang, M.T.; Ho, C.T.; Lee, C.Y. *Phenolic Compounds in Food and their Effects on Health: Volume II: Antioxidants and Cancer Prevention.* Washington, DC: American Chemical Society, 1992, 506, 102–117.

Ishimoto, K.; Hatanaka, N.; Otani, S.; Maeda, S.; Xu, B.; Yasugi, M.; Moore, J.E.; Suzuki, M.; Nakagawa, S.; Yamasaki, S. Tea crude extracts effectively inactivate severe acute respiratory syndrome coronavirus 2. *Lett. Appl. Microbiol.* 2021, 74(1), 2–7.

Javanmardi, J.; Stushnoff, C.; Locke, E.; Vivanco, J.M. Antioxidant activity and total phenolic content of Iranian Ocimum acces-sions. *J. Food Chem.* 2003, 83, 547–550.

Ji, B.U.; Chow, W.H.; Hsing, A.W. Green tea consumption and the risk of pancreatic and colorectal cancers. *Int. J. Cancer* 1997, 70(3), 255–258.

Jing, Y.; Han, G.; Hu, Y.; Bi, Y.; Li, L.; Zhu, D. Tea consumption and risk of type 2 diabetes: A meta-analysis of cohort studies. *J. Gen. Intern. Med.* 2009, 24(5), 557–562.

Johnson, I.; Williamson, G.; *Phytochemical Functional Foods.* Cambridge: CRC, 2003.

Kampa, M.; Nifli, A.-P.; Notas, G.; Castanas, E. Polyphenols and can-cer cell growth. *J. Rev. Physiol. Biochem. Pharmacol.* 2007, 159, 79–113.

Kanis, J.; Johnell, O.; Gullberg, B. Risk factors for hip fracture in men from southern Europe: The MEDOS study. Mediterranean osteoporosis study. *J. Osteoporos Int.* 1999, 9(1), 45–54.

Kao, Y.H.; Hiipakka, R.A.; Liao, S.; Modulation of endocrine sys-tems and food intake by green tea epigallocatechin gallate. *Endocrinology,* 2000, 141, 980–987.

Katiyar, S.K.; Mukhtar H. Tea in chemoprevention of cancer: epidemiological and experimental studies. *Int. J. Oncol.* 1996, 8(2), 221–238.

Keflie, T.S.; Biesalski, H.K. Micronutrients and bioactive substances: Their potential roles in combating COVID-19. *J. Nutr.* 2021, 84, 111103.

Keller, A.; Wallace, T.C. Tea intake and cardiovascular disease: an umbrella review. *J. Ann. Med.* 2021, 53(1), 929–944.

Khan, F.; Bashir, A.; Al-Mughairbi, F. Purple tea composition and inhibitory effect of anthocyanin-rich extract on cancer cell proliferation. *J. Med. Aromat. Plants* 2018, 7(06), 322–326.

Kim, H.S.; Kim, M.H.; Jeong, M.; Hwang, Y.S.; Lim, S.H.; Shin, B.A.; Ahn, B.W.; Jung, Y.D. EGCG blocks tumor promoter-induced MMP-9 expression via suppression of MAPK and AP-1 activation in human gastric AGS cells. *J. Anticancer Res.* 2004, 24, 747–753.

Klein, R.D.; Fischer, S.M.; Black tea polyphenols inhibit IGF-I-induced signaling through AKT in normal prostate epithelial cells and Du145 prostate. *Carcinogenesis*, 2002, 23, 217–221.

Koch, W.; Kukula-Koch, W.; Komsta, Ł.; Marzec, Z.; Szwerc, W.; Głowniak, K. Green tea quality evaluation based on its catechins and metals composition in combination with chemometric analysis. *J. Mol.* 2018, 23(7), 1689.

Kuroda, Y.; Hara, Y. *Health Effects of Tea and its Catechins*. Springer, 2004.

Lange, K. Tea in cardiovascular health and disease: a critical appraisal of the evidence. *J. Food Sci. Hum. Wellness* 2022, 11(3), 445–454.

Le Marchand, L. Cancer preventive effects of flavonoids-A review. *J. Biomed. Pharmacother* 2002, 56(6), 296–301.

Lee, K.I.; Kundu, J.K.; Kim, S.O.; Chun, K-S.; Lee, H.J.; Surh, Y-J. Cocoa polyphenols inhibit phorbol ester-induced superoxide anion formation in cultured HL-60 cells and expression of cyclooxygenase-2 and activation of NF- Kappab and Mapks in mouse skin in vivo. *J. Nutr.* 2006, 136(5), 1150–1155.

Lee, M.J.; Lambert, J.D.; Prabhu, S.; Meng, X.; Lu, H.; Maliakal, P.; Ho, C.T.; Yang, C.S. Delivery of tea polyphenols to the oral cavity by green tea leaves and black tea extract. *J. Cancer Epidemiol. Biomarkers Prev.* 2004, 13(1), 132–137.

Lee, J.H.; Park, J.; Shin, D.W. The molecular mechanism of polyphenols with anti-aging activity in aged human dermal fibroblasts. *J. Mol.* 2022, 27, 4351.

Levites, Y.; Amit, T.; Youdim, M.B.H.; Mandel, S. Involvement of protein kinase C activation and cell survival/ cell cycle genes in green tea polyphenol (–)- Epigallocatechin 3-gallate neuroprotective action. *J. Biol. Chem.* 2002, 277(34), 30574–30580.

Liao, S.; Kao, Y.H.; Hiipakka, R.A. Green tea: Biochemical and biological basis for health benefits. *J. Vitam. Horm.* 2001, 62, 1–94.

Lin, L.Z.; Harnly, J.M.A Screening method for the identification of glycosylated flavonoids and other phenolic compounds using a standard analytical approach for all plant materials. *J. Agric. Food Chem.* 2007, 55(4), 1084–1096.

Lin, J.-K.; Liang, Y.-C.; Lin-Shiau, S.-Y. Cancer chemoprevention by tea polyphenols through mitotic signal transduction blockade. *J. Biochem. Pharmacol.* 1999, 58(6), 911–915.

Liu, Z.; Hu, M. Natural Polyphenol Disposition via coupled metabolic pathways. *J. Expert Opin. Drug Metab. Toxicol.* 2007, 3(3), 389–406.

Lung, J.; Lin, Y.S.; Yang, Y.H.; Chou, Y.L.; Chang, G.H.; Tsai, M.S.; Hsu, C.M.; Yeh, R.A.; Shu, L.H.; Cheng, Y.C.; Liu, H. T; Wu, C.Y. The potential SARS-CoV-2 entry inhibitor. *J. bioRxiv* 2020.

Maiti, S.; Banerjee, A. Epigallocatechin gallate and theaflavin gallate interaction in SARS-CoV-2 spike-protein central channel with reference to the hydroxychloroquine interaction: Bioinformatics and molecular docking study. *J. Drug Dev. Res.* 2021, 82(1), 86–96.

Manach, C.; Scalbert, A.; Morand, C.; Rémésy, C.; Jiménez, L. Polyphenols: Food sources and bioavailability. *Am. J. Clin. Nutr.* 2004, 79, 727–747

March, R.E.; Lewars, E.G.; Stadey, C.J.; Miao, X.-S.; Zhao, X.; Metcalfe, C.D. A comparison of flavonoid glycosides by electrospray tandem mass spectrometry. *Int. J. Mass Spectrom.* 2006, 248(1–2), 61–85.

March, R.E.; Miao, X.S.A Fragmentation study of kaempferol using electrospray quadrupole time-of-flight mass spectrometry at high mass resolution. *Int. J. Mass Spectrom.* 2004, 231(2–3), 157–167.

Martinez-Gomez, A.; Caballero, I.; Blanco, C.A. Phenols and melanoidins as natural antioxidants in beer. Structure, reactivity and antioxidant activity. *J. Biomol.* 2020, 10(3), 400.

Masella, R.; Dibenedetto, R.; Vari, R.; Filesi, C.; Giovannini, C. Novel mechanisms of natural antioxidant compounds in biological systems: Involvement of glutathione and glutathione-related enzymes. *J. Nutr. Biochem.* 2005, 16(10), 577–586.

Mathew, A.G.; Parpia, H.A.B. Food browning as a polyphenol reaction. *J. Adv. Food Res.* 1971, 19, 75–145.

Mayer, A.M. Polyphenol oxidases in plants and fungi: Going places? A review. *Phytochem.* 2006, 67(21), 2318–2331.

Mazza, G.J. Anthocyanins and heart health. *J. Ann. Ist. Super Sanita.* 2007, 43(4), 369–374.

Meng, J.M.; Cao, S.Y.; Wei, X.L.; Gan, R.Y.; Wang, Y.F.; Cai, S.X.; Xu, X.Y.; Zhang, P.Z.; Li, H.B. Effects and mechanisms of tea for the prevention and management of diabetes mellitus and diabetic complications: An updated review. *J. Antioxidants* 2019, 8(6), 170.

Mercer, L.D.; Kelly, B.L.; Horne, M.K.; Beart, P.M. Dietary polyphenols protect dopamine neurons from oxidative insults and apoptosis: Investigations in primary rat mesencephalic cultures. *J. Biochem. Pharmacol.* 2005, 69(2), 339–345.

Merken, H.M.; Beecher, G.R. Measurement of food flavonoids by high- performance liquid chromatography: A review. *J. Agric. Food Chem.* 2000, 48(3), 577–599.

Mhatre, S.; Naik, S.; Patravale, V. A molecular docking study of EGCG and theaflavin digallate with the druggable targets of SARS-CoV-2. *J. Comput. Biol. Med.* 2021, 129, 104137.

Mhatre, S.; Srivastava, T.; Naik, S.; Patravale, V. Antiviral activity of green tea and black tea polyphenols in prophylaxis and treatment of COVID-19. *J. Phytomedicine*, 2021, 85,153286.

Mieczan, W.A.; Tomaszewska, E.; Jachimowicz, K. Antioxidant, anti-inflammatory, and immunomodulatory properties of tea—The positive impact of tea consumption on patients with autoimmune diabetes. *J. Nutr.* 2021, 13(11), 3972.

Mohamed,G.A.; Ibrahim, S. R.; Elkhayat,M. S.; El Dine, R.S. Natural anti-obesity agents. *Bulletin of Faculty of Pharmacy, Cairo University*, Elsevier B.V. on behalf of Faculty of Pharmacy, Cairo University, 2014, 52(2), 269–284

Montone, C.M.; Aita, S.E.; Arnoldi, A.; Capriotti, A.L.; Cavaliere, C.; Cerrato, A.; Lammi, C.; Piovesana, S.; Ranaldi, G.; Lagana, A. Characterization of the trans-epithelial transport of green tea (C. sinensis) catechin extracts with in vitro inhibitory effect against the SARS-CoV-2 papain-like protease activity. *J. Mol.* 2021, 26(21), 6744.

Montoro, P.; Tuberoso, C.I.G.; Perrone, A.; Piacente, S.; Cabras, P.; Pizza, C. Characterisation by liquid chromatography-electrospray tandem mass spectrometry of anthocyanins in extracts of myrtus communis L. berries used for the preparation of myrtle liqueur. *J. Chromatogr. A*, 2006, 1112(1–2), 232–240.

Mostrom, M.; Evans, T.J. *Chapter 52 – Phytoestrogens in Reproductive and Development Toxicology*. Academic Press. 2011, 707–722. ISBN: 9780123820327. https://doi.org/10.1016/B978-0-12-382032-7.10052-9

Mukhtar, H.; Ahmad, N.; Cancer chemoprevention: future holds in multiple agents. *Toxicol. Appl. Pharmacol.* 1992, 158, 207–210.

Munin, A.; Edwards-Lévy, F. Encapsulation of natural polyphenolic compounds; A review. *J. Pharmaceutics* 2011, 3(4), 793–829.

Nagata, T. New analytical methods for studying tea quality components. *Tea Res. J.* 1990, 72, 53.

Nakabayashi, T; Ina, K.; Sakata, K.; Shuppan, K. Chemical components in tea leaves, in chemistry and function of green tea, black tea, and oolong tea. *Kawasaki, Japan*, 1991, 20, 3–39.

Nakachi, K.; Suemasu, K.; Suga, K.; Takeo, T.; Imai, K.; Higashi, Y. Influence of drinking green tea on breast cancer malignancy among Japanense patients. *Jpn. J. Cancer Res.* 1998, 89(3), 254–261.

Nicolas, J.; Rouet-Mayer, M.A. *Encyclopedia of Food Sciences and Nutrition, Polyphenol Oxidases (EC1.10.3.1)*, London: Academic press, 2003. ISBN: 978-0122270550.

Nijveldt, R.J.; Van Nood, E.; Van Hoorn, D.E.; Boelens, P.G.; Van Norren, K.; Van Leeuwen, P.A. Flavonoids: A review of probable mechanisms of action and potential applications. *Am. J. Clin. Nutr.* 2001, 74(4), 418–425.

Nomura, M.; Kaji, A.; He, Z.; Ma, W.Y.; Miyamoto, K.; Yang, C.S.; Dong, Z. Inhibitory mechanisms of tea polyphenols on the ultraviolet B-activated phosphatidylinositol 3-kinase-dependent pathway. *J. Biol. Chem.* 2001, 276(49), 46624–46631.

Ofodile, O.N.F.C. Cardiovascular disease could be contained based on currently available data! *J. Dose Response* 2006, 4(3), 225–254.

Oguni, I.; Nasu, K.; Kanaya, S.; Ota, Y.; Yamamoto, S.; Nomura, T. Epidemiological and experimental studies on the anti-tumor activity by green tea extracts. *Jpn. J. Nutr.* 1989, 47, 93–102.

Okabe, S.; Suganuma, M.; Hayashi, M.; Sueoka, E.; Komori, A.; Fujiki, H. Mechanisms of growth inhibition of human lung cancer cell line, pc-9, by tea polyphenols. *Jpn. J. Cancer Res.* 1997, 88(7), 639–643.

Ostrowska, J. Are teas the universal antioxidants? In: *Leading Edge Antioxidants Research*. H. V. Panglossi (ed.), New York: Nova Science Publishers, 2007, 89–144.

Ovaskainen, M.L.; Törrönen, R.; Koponen, J.M.; Sinkko, H.; Hellström, J.; Reinivuo, H.; Mattila, P. Dietary intake and major food sources of polyphenols in finnish adults. *J. Nutr.* 2008, 138(3), 562–6.

Owuor, P.O.; Othieno, C.O. Changes in chemical composition of black tea due to pruning. *J. Tropical Sci.* 1988, 28(2), 127–132.

Pantel, S.; Adams, S.; Lee, L. Inhibition of herpes simplex virus-1 by the modified green tea polyphenol EGCG-stearate. *J. Adv. Biosci. Biotechnol.* 2018, 9, 679–690.

Paredes, A.; Alzuru, M.; Mendez, J.; Rodríguez-Ortega, M. Anti-sindbis activity of flavanones hesperetin and naringenin. *J. Biol. Pharm Bull.* 2003, 26(1), 108–109.

Paschka, A.G.; Butler, R.; Young, C.Y. Induction of apoptosis in prostate cancer cell lines by the green tea component, (-)-epigallocatechin-3-gallate. *J. Cancer Lett.* 1998, 130(1–2), 1–7.

Patel, A. Estimation of flavonoid, polyphenolic content and in-vitro antioxidant capacity of leaves of tephrosia purpurea linn. *Int. J. Pharm Sci. Res.* 2010, 1(1), 66–77.

Peele, K.A.; Durthi, C.P.; Srihansa, T.; Krupanidhi, S.; Ayyagari, V.S.; Babu, D.J.; Indira, M.; Reddy, A.R.; Venkateswarulu, T.C. Molecular docking and dynamic simulations for antiviral compounds against SARS-CoV-2: A computational study. *J. Inform Med. Unlocked* 2020, 19, 100345.

Perron, N.R.; Brumaghim, J.L. A review of the antioxidant mechanisms of polyphenol compounds related to iron binding. *J. Cell Biochem. Biophys.* 2009, 53(2), 75–100.

Polidori, M.C. Oxidative stress and risk factors for Alzheimer's disease: Clues to prevention and therapy. *J. Alzheimers Dis.* 2004, 6(2), 185–91.

Prasad, K.N.; Cole, W.C.; Kumar, B. Multiple antioxidants in the prevention and treatment of Parkinson's disease. *J. Am. Coll. Nutr.* 1999, 18(5), 413–423.

Rawangkan, A.; Kengkla, K.; Kanchanasurakit, S.; Duangjai, A.; Saokaew, S. Anti-influenza with green tea catechins: A systematic review and meta-analysis. *J. Mol.* 2021, 26(13), 4014.

Rice-Evans, C.; Miller, N.; Paganga, G. Antioxidant properties of phenolic compounds. *J. Trends Plant Sci.* 1997, 2(4), 152–159.

Roberts, E.A.H. Oxidative condensation of flavonols in tea fermentation. *Chem. & Ind. (London)* 1957, p. 1355.

Roberts, E.A.H. Interaction of flavonol orthoquinones with cysteine and glutathione. *Chem. & Ind. (London)* 1959, p. 995.

Sakakibara, H.; Honda, Y.; Nakagawa, S.; Ashida, H.; Kanazawa, K. Simultaneous determi nation of all polyphenols in vegetables, fruits, and teas. *J. Agric. Food Chem.* 2003, 51(3), 571–581.

Samarghandian, S.; Azimi-Nezhad, M.; Farkhondeh, T. Catechin treatment ameliorates iabetes and its complications in strep-tozotocin-induced diabetic rats. *J. Dose Response* 2017, 15(1), 1559325817691158.

Sánchez-Moreno, C.; A. Larrauri, J.; Saura-Calixto, F. Free radical scavenging capacity and inhibition of lipid oxidation of wines, grape juices and related polyphenolic constituents. *J. Food Res. Int.* 1999, 32, 407–412.

Saric, S.; Sivamani, R.K. Polyphenols and sunburn. *Int. J. Mol. Sci.* 2016, 17(9), 1521.

Scalbert, A.; Williamson, G. Dietary intake and bioavailability of polyphenols. *J. Nutr.* 2000, 130(8SSuppl.), 2073S–2085S.

Schroeter, H.; Boyd, C.; Spencer, J.P.E.; Williams, R.J.; Cadenas, E.; Rice- Evans, C. MAPK Signaling in neurodegeneration: Influences of flavonoids and of nitric oxide. *J. Neurobiol. Aging* 2002, 23(5), 861–880.

Schroeter, H.; Spencer, J.P.; Rice-Evans, C.; Williams, R.J. Flavonoids protect neurons from oxidized low-density-lipoprotein-induced apoptosis involving C Jun N Terminal Kinase (JNK), C-Jun and Caspase-3. *Biochem. J.* 2001, 358(3), 547–557.

Sies, H. Oxidative stress: Oxidants and antioxidants. *J. Exp. Physiol.* 1997, 82(2), 291–295.

Singh, M.; Arseneault, M.; Sanderson, T.; Murthy, V.; Ramassamy, C. Challenges for research on polyphenols from foods in Alzheimer's disease: Bioavailability, metabolism, and cellular and molecular mechanisms. *J. Agric. Food Chem.* 2008, 56(13), 4855–4873.

Singh, S.; Sk, M.F.; Sonawane, A.; Kar, P.; Sadhukhan, S. Plant-derived natural polyphenols as potential antiviral drugs against SARS-CoV-2 via RNA-dependent RNA polymerase (RdRp) inhi-bition: an in-silico analysis. *J. Biomol. Struct. Dyn.* 2021, 39(16), 6249–6264.

Singh, S.; Pandey, R.; Tomar, S.; Varshney, R.; Sharma, D.; Gangenahalli, G. A brief molecular insight of COVID19: Epidemiology, clinical manifestation, molecular mechanism, cellular tropism and immunopathogenesis. *J. Mol. Cell Biochem.* 2021, 476(11), 3987–4002.

Sirotkin, A.V.; Kolesárová, A. The anti-obesity and health-promoting effects of tea and coffee. *J. Physiol. Res.* 2021, 70(2), 161–168.

Skrypnik, K.; Suliburska, J.; Skrypnik, D.; Pilarski, Ł.; Reguła, J.; Bogdański, P. The genetic basis of obesity complications. *J. Acta scientiarum polonorum Technologia alimentaria,* 2017,16(1), 83–91

Song, J.M.; Lee, K.H.; Seong, B.L. Antiviral effect of catechins in green tea on influenza virus. *J. Antiviral Res.* 2005, 68(2), 66–74.

Soong, Y.Y.; Barlow, P.J. Isolation and structure elucidation of phenolic compounds from longan (dimocarpus longan lour.) seed by high-performance liquid chromatography-electrospray ionization mass spectrometry. *J. Chromatogr. A.* 2005, 1085(2), 270–207.

Speirs, V. Phytoestrogens. In: *Encyclopedia of Cancer.* M. Schwab (ed.), Berlin, Heidelberg: Springer. 2008, 2886–2887.

Spencer, J.P.E.; Kuhnle, G.G.C.; Williams, R.J.; Rice-Evans, C. Intracellular metabolism and bioactivity of quercetin and its in vivo metabolites. *Biochem. J.* 2003, 372(pt 1), 173–181.

Su, X.; Duan, J.; Jiang, Y.; Duan, X.; Chen, F. Polyphenolic profile and antioxidant activities of oolong tea infusion under various steeping conditions. *Int. J. Mass Spectrom.* 2007, 8(12), 1196–1205.

Subissi, L.; Posthuma, C.C.; Collet, A.; Zevenhoven-Dobbe, J.C.; Gorbalenya, A.E.; Decroly, E.; Snijder, E.J.; Canard, B.; Imbert, I. One severe acute respiratory syndrome coronavirus protein complex integrates processive RNA polymerase and exonuclease activities. *Proc. Natl. Acad. Sci. U.S.A.* 2014, 111(37), E3900–E3909.

Sun, K.; Wang, L.; Ma, Q.; Cui, Q.; Lv, Q.; Zhang, W.; Li, X. Association between tea consumption and osteoporosis: A meta-analysis. *J. Med. (Baltimore)* 2017, 96(49), e9034.

Sung, K. C.; Lee, M. Y.; Kim, Y. H.; Huh, J. H.; Kim, J. Y.; Wild, S. H.; Byrne, C. D. Obesity and incidence of diabetes: Effect of absence of metabolic syndrome, insulin resistance, inflammation and fatty liver. *J.Atherosclerosis,* 2018, 275, 50-57.

Takeda, Y.; Tamura, K.; Jamsransuren, D.; Matsuda, S.; Ogawa, H. Severe acute respiratory syndrome coronavirus-2 inactivation activity of the polyphenol-rich tea leaf extract with concentrated theaflavins and other virucidal catechins. *J. Mol.* 2021, 26(16), 4803.

Takeo, T. Green and semi-fermented teas. In: *Tea: Cultivation to consumption.* K. C. Willson and M. N. Clifford (eds.), London: Chapman & Hall, 1992, 413–457.

Tanaka, T.; Watarumi, S.; Matsuo, Y.; Kamei, M.; Kouno, I. Production of theasinensins A and D, epigallocatechin gallate dimers of black tea, by oxidation-reduction dismutation of dehydrotheasinensin A. *J. Tetrahedron* 2003, 59(40), 7939–7947.

Tapiero, H.; Tew, K.D.; Nguyen Ba, G.; Mathé, G. Polyphenols: Do they play a role in the prevention of human pathologies? *J. Biomed. Pharmacother.* 2002, 56(4), 200–207.

Thichanpiang, P.; Wongprasert, K. Green tea polyphenol epigallocatechin-3-gallate attenuates TNF-α-induced intercellular adhesion molecule-1 expression and monocyte adhesion to retinal pigment epithelial cells. *Am. J. Chin Med.* 2015, 43(1), 103–119.

Tsuchiya, H.; Sato, M.; Kato, H.; Okubo, T.; Juneja, L. R.; Kim, M. Simultaneous determination of catechins in human saliva by high-performance liquid chromatography. *J. Chromatogr. B Biomed. Sci. App.* 1997, 703(1–2), 253–258.

Urquiaga, I.; Leighton, F. Plant polyphenol antioxidants and oxidative stress. *J. Biol. Res.* 2000, 33(2), 55–64.

Vardhan, S.; Sahoo, S.K. Virtual screening by targeting proteolytic sites of furin and TMPRSS2 to propose potential compounds obstructing the entry of SARS-CoV-2 virus into human host cells. *J. Tradit. Complement Med.* 2022, 12(1), 6–15.

Vasto, S.; Barera, A.; Rizzo, C.; Di Carlo, M.; Caruso, C.; Panotopoulos, G. Mediterranean diet and longevity: An example of nutraceuticals? *J. Curr. Vasc. Pharmacol.* 2014, 12(5), 735–738.

Vázquez-Calvo, A.; Oya, N.; Acebes, M.A.M.; Moruno, E.G.; Saiz, J.C. Antiviral properties of the natural polyphenols delphinidin and epigallocatechin gallate against the flaviviruses west nile virus, zika virus, and dengue virus. *J. Front Microbiol.* 2017, 8, 1314.

Velayutham, P.; Babu, A; Liu, D. Green tea catechins and cardiovascular health. *J. Curr. Med. Chem.* 2008; 15(18), 1840–1850.

Vincent, A.; Fitzpatrick, L.A. Soy isoflavones: Are they useful in menopause? *J. Mayo Clin. Proc.* 2000, 75(11), 1174–1184.

Visioli, F.; Galli, C. The role of antioxidants in the mediterranean diet. *J. Lipids* 2001, 36, S49–S52

Wang, X.; Dong.W; Zhang.X; Zhu.Z; Chen.Y; Liu.X; Guo.C. Antiviral mechanism of tea polyphenols against porcine reproductive and respiratory syndrome virus. *J. Pathogens* 2021, 10(2), 202.

Wang, S;Moussa.M.N; Chen.L; Mo.H; Shastri.A; Su.R; Bapat, P; Kwun.I; Shen, L.C. Novel insights of dietary polyphenols and obesity. *J. Nutr. Biochem.* 2014, 25(1), 1–18.

Wang, Y.; Li, S.;Zheng, X.; Lu, J.; Liang, Y. Antiviral effects of green tea EGCG and Its potential application against COVID-19. *J. Mol.* 2021, 26(13), 3962.

WHO. World health organization (www.who.int).

Williams, R.J.; Spencer, J.P.E.; Rice-Evans, C. Flavonoids: Antioxidants or signalling molecules? *J. Free Radic. Biol. Med.* 2004, 36(7), 838–349.

Williamson, G.; Manach, C. Bioavailability and bioefficacy of polyphenols in humans. II. Review of 93 intervention studies. *Am. J. Clin. Nutr.* 2005, 81 (1Suppl), 243S–255S.

Winiarska-Mieczan, A. Protective effect of tea against lead and cadmium-induced oxidative stress-a review. *J.Biometals.* 2018, 1(6), 909–926.

Wiseman, H.; Halliwell, B. Damage to DNA by reactive oxygen and nitrogen species: Role in inflammatory disease and progression to cancer. *Biochem. J.* 1996, 313, 17–29.

Wright, L.P.; *Biochemical Analysis for Identification of Quality in Black Tea (Camellia Sinensis).* University of Pretoria (ETD), 2005.

Yamada, J.; Tonita, Y. Antimutagenic activity of water extract of black tea and oolong tea. *J. Biosc. Biotect. Biochem.* 1994, 58(12), 2197–2200.

Yamamoto, T.; Juneja, L.R.; Chu, D.C.; Kim, M. *Chemistry and Applications of Green Tea.* Panama City, FL: CRC, 1997.

Yang, C.; Du, W.; Yang, D. Inhibition of Green Tea Polyphenol EGCG ((−)-Epigallocatechin-3-Gallate) on the Proliferation of Gastric Cancer Cells by Suppressing Canonical Wnt/β-Catenin Signalling Pathway. *Int. J. Food Sci. Nutr.* 2016, 67(7), 818–827.

Yang, C.S.; Ju, J.; Lu, G.; Xiao, H.; Hao, X.; Sang, S.; Lambert, J.D. Cancer prevention by tea and tea polyphenols. *Asia Pac. J. Clin. Nutr.* 2008, 17, 245–248.

Yang, G.Y.; Liao, J.; Kim, K.; Yurkow, E.J.; Yang, C.S. Inhibition of growth and induction of apoptosis in human cell lines by tea polyphenols. *J. Carcinog.* 1998, 19(4), 611–616.

Yang, C.S.; Wang, Z.Y.; Tea and cancer. *J. Natl. Cancer Inst.* 1993, 85(13), 1038–1049.

Yang, C.S.; Yang, G.Y.; Landau, J.M.; Kim, S.; Liao, J. Tea and tea polyphenols inhibit cell hyperproliferation, lung tumorigenesis, and tumor progression. *J. Exp. Lung Res.* 1998, 24(4), 629–639.

Yang, Z.F.; Bai, L.P.; Huang, W.B.; Li, X.Z.; Zhao, S.S.; Zhong, N.S.; Jiang, Z.H. Comparison of in vitro antiviral activity of tea polyphenols against influenza A and B viruses and structure–activity relationship analysis. *J. Fitoterapia* 2014, 93, 47–53.

Yoshioka, H.; Akai, G.; Yoshinga, K.; Hasegewa, K.; Yoshioka, H. Protecting effect of a green tea percolate and its main constituents against gamma ray-induced scission of DNA. *J. Biosci. Biotechnol. Biochem.* 1996, 60(1), 117–119.

Zern, T.L.; Wood, R.J.; Greene, C.; West, K.L.; Liu, Y.; Aggarwal, D.; Shachter, N.S.; Fernandez, M.L. Grape polyphenols exert a cardioprotective effect in pre- and postmenopausal women by lowering plasma lipids and reducing oxidative stress. *J. Nutr.* 2005, 135(8), 1911–1917.

Zhang, G.; Miura, Y.; Yagasaki, K. Induction of apoptosis and cell cycle arrest in cancer cells by in vivo metabolites of teas. *J. Nutr. Cancer* 2000, 38(2), 265–273.

Zhai, Y.; Sun, F.; Li, X.; Pang, H.; Xu, X.; Bartlam, M.; Rao, Z. Insights into SARS-CoV transcription and replication from the structure of the nsp7-nsp8 hexadecamer. *J. Nat. Struct. Mol. Biol.* 2005, 12(11), 980–986.

Zhao, M.; Yu, Y.; Sun, L.M.; Xing, J.Q.; Li, T.; Zhu, Y.; Wang, M.; Yu, Y.; Xue, W.; Xia, T.; Cia, H.; Han, Q.Y.; Yin, X.; Li, W.H.; Li, A.L.; Cui, J.; Yuan, Z.; Zhang, R.; Zhou, T.; Zhang, X.M.; Li, T. GCG inhibits SARS-CoV-2 replication by disrupting the liquid phase condensation of its nucleocapsid protein. *J. Nat. Commun.* 2021, 12(1), 2114.

Zhang, J.; Shen, X.; Yan, Y.; Wang, Y.; Cheng, Y. Discovery of anti-SARS-CoV-2 agents from commercially available flavor via docking screening. 2020.

Zhong, Y.; Ma, N.; Shahidi, F. Antioxidant and antiviral activities of lipophilic epigallocatechin gallate (EGCG) derivatives. *J. Funct. Foods* 2011, 4(1), 87–93.

Zumbé, A. Polyphenols in cocoa: Are there health benefits? *J. Nutr. Bulletin* 1998, 23, 94–102.

Compound Index

Note: **Bold** page numbers refer to tables and *italic* page numbers refer to figures.

General Index

Note: **Bold** page numbers refer to tables and *italic* page numbers refer to figures.

109